Creative DIY Microcontroller Projects with TinyGo and WebAssembly

A practical guide to building embedded applications for low-powered devices, IoT, and home automation

Tobias Theel

BIRMINGHAM—MUMBAI

Creative DIY Microcontroller Projects with TinyGo and WebAssembly

Copyright © 2021 Packt Publishing

Group Product Manager: Aaron Lazar

Publishing Product Manager: Alok Dhuri

Senior Editor: Nitee Shetty

Content Development Editor: Tiksha Lad

Technical Editor: Karan Solanki

Copy Editor: Safis Editing

Project Coordinator: Deeksha Thakkar

Proofreader: Safis Editing

Indexer: Manju Arasan

Production Designer: Vijay Kamble

First published: April 2021

Production reference: 1160421

Published by Packt Publishing Ltd.

Livery Place

35 Livery Street

Birmingham

B3 2PB, UK.

ISBN 978-1-80056-020-8

www.packt.com

To Sir Patrick Stewart, for being a shining light of humanity even in the darkest days. And to my beloved dog, Luca, for silently accepting that some walks have been shorter than usual while writing this book.

– Tobias Theel

Contributors

About the author

Tobias Theel works as the Technical Lead and DevOps for a German FinTech startup fino and since 2020 he has also started working for RegTech startup, ClariLab, as Lead Software Engineer. Being a software architect and an expert for Go and TinyGo alongside C# and Java, he is also iSAQB certified. Theel is a highly enthusiastic community contributor and is among the top 10% responders in C# and Unity3D as well as top 20% responders in .NET, Go, and Visual Studio on StackOverflow.

When not programming for fino or ClariLab, he can be found developing games, mainly at game jams such as the Ludum Dare Jam, where he develops games from scratch within 72 hours. As an active speaker at tech talks and a participant for numerous hackathons, Theel loves to share his knowledge of software development with fellow enthusiasts.

I want to thank my Team at ClariLab for granting me so many days off in order to finish this book. Also special thanks to Johannes Kolata, for giving so much valuable input for the book.

About the reviewers

Enrico von Otte started to learn programming as an 11-year-old child on the good old Commodore C64. Later he coded on an Amiga 2000 and in the mid 90s finally on PC. He is working in professional software development from 2005. After developing hardware close to software in C and C++, he moved to the C# world in 2008.

He developed GIS systems and document management systems until 2015. Now he is a professional software architect with a strong affinity to self-made code and the motto "Assembling is not developing".

Johannes Kolata started coding 8 years ago, working on document management systems. Since 2018, Kolata works as a Software Backend Engineer for German FinTech fino, which is best known for building the first digital bank account switch. Along the way, he has become an expert in developing in C#, Java, and Golang. He also became adept in C++ and has gained insights in multiple other programming languages. Besides working in FinTech industry, he is a passionate open source and 3D-printing enthusiast and has been building home automation systems from scratch that feature CAN communication, go API, and an Angular dashboard. When not working on projects, Kolata participates in game jams and hackathons.

Table of Contents

3

Building a Safety Lock Using a Keypad

4

Building a Plant Watering System

8

Automating and Monitoring Your Home through the TinyGo Wasm Dashboard

Appendix – "Go"ing Ahead

Assessments

Afterword

Other Books You May Enjoy

Index

Preface

If JavaScript or C# can run on microcontrollers, then Go can do it even better. While standard Go produces huge binaries, TinyGo produces binaries that fit on the smallest devices. Why should you choose Go for microcontroller and **Wasm** (short for **WebAssembly**) programming? My favorite reasons are that Go is easy to learn, easy to read, and easy to write. Also, Go comes with a powerful standard library that is loosely coupled and has strong concurrency capabilities included.

If you love Go as a language, then this book is for you. After working through this book, you'll have all the tools and knowledge you need to build all the microcontroller projects that you have ever dreamed of. Plus, as a further benefit, you will be able to build dashboards and home control apps using Wasm for your home automation projects. All of this can be achieved using TinyGo.

If you have never worked with microcontrollers before, here are some reasons why you should try it out:

- If you already are a programmer, it is a cool thing to see code affecting real-world appliances. It really is a great feeling to complete a project and finally see the motor turning, the LED blink, the buzzer beep, and so on.

- You will be constantly learning new things and gaining a deeper understanding of how computers work in general, as you will become familiar with different types of bus systems, protocols, hardware interfaces, and much more.

- The possibilities are virtually limitless when you're playing with microcontrollers. You are not bound to what is available on the market, as you can simply build everything on your own.

- You can learn how to write small, efficient programs to tell a microcontroller what you want from it. This will also help you become a better developer in general.

- You can contribute to cool projects and get in touch with great communities of like-minded people.

Who this book is for

If you are a Go developer who wants to program low-powered devices and hardware such as Arduino UNO and Arduino Nano IoT 33, or if you are a Go developer who wants to extend your knowledge of using Go with WebAssembly while programming Go in the browser, then this book is for you. Go hobbyist programmers who are interested in learning more about TinyGo by working through DIY projects will also find this hands-on guide useful.

What this book covers

Chapter 1, Getting Started with TinyGo, sees you set up TinyGo and compile your first program!

Chapter 2, Building a Traffic Lights Control System, has you build a traffic lights control system, including pedestrian lights and a button; you will learn how to make use of Goroutines in TinyGo.

Chapter 3, Building a Safety Lock Using a Keypad, looks at making use of a 4x4 keypad and a servomotor to build a lock that opens when the correct passcode has been entered.

Chapter 4, Building a Plant Watering System, explains how to use different types of sensors to build an automatic plant watering system, so you do not have to water your plants manually anymore!

Chapter 5, Building a Touchless Handwash Timer, explores using a four-digit, seven-segment display and an ultrasonic distance sensor to recognize the movement of a nearby object to start a timer that will tell us when we have washed our hands for long enough.

Chapter 6, Building Displays for Communication Using I2C and SPI Interfaces, explains the concepts of **Inter-Integrated Circuit** (**I2C**) and **Serial Peripheral Interface** (**SPI**) by having you use displays that communicate using I2C and SPI buses. By the end of the chapter, you will know how to use different types of displays in TinyGo.

Chapter 7, Displaying Weather Alerts on the TinyGo Wasm Dashboard, is where you will build and serve a Wasm application that displays sensor data sent from an Arduino Nano 33 IoT over Wi-Fi.

Chapter 8, Automating and Monitoring Your Home through a TinyGo Wasm Dashboard, explains how to control and monitor devices in your home using a Wasm dashboard.

Chapter 9, Appendix–"Go"ing Ahead, covers some concepts that did not find their way into the previous chapters.

To get the most out of this book

All code examples have been tested with Go 1.16.2 on Ubuntu, but they will also work with future releases of Go and other operating systems. Visual Studio Code has been used as the editor throughout the book, but any other editor can be used.

Software/hardware covered in the book	Operating system requirements
Go 1.15.x or newer	Windows, macOS X, or Linux
Visual Studio Code	Windows, macOS X, or Linux

If you are using the digital version of this book, we advise you to type the code yourself or access the code via the GitHub repository (link available in the next section). Doing so will help you avoid any potential errors related to the copying and pasting of code.

I would love to see the projects that you build after reading this book on social media. Feel free to tag me on Twitter using the following tag: @Nooby_Games.

Download the example code files

You can download the example code files for this book from GitHub at https://github.com/PacktPublishing/Creative-DIY-Microcontroller-Projects-with-TinyGo-and-WebAssembly. In case there's an update to the code, it will be updated on the existing GitHub repository.

We also have other code bundles from our rich catalog of books and videos available at https://github.com/PacktPublishing/. Check them out!

Code in Action

Code in Action videos for this book can be viewed at https://bit.ly/3cYZOh4.

Download the color images

We also provide a PDF file that has color images of the screenshots/diagrams used in this book. You can download it here: https://static.packt-cdn.com/downloads/9781800560208_ColorImages.pdf.

Conventions used

There are a number of text conventions used throughout this book.

`Code in text`: Indicates code words in text, database table names, folder names, filenames, file extensions, pathnames, dummy URLs, user input, and Twitter handles. Here is an example: "When we receive the start of the command, we append all subsequent characters to `commandBuffer`."

A block of code is set as follows:

```
data, err := uart.ReadByte()
if err != nil {
    println(err.Error())
}
```

When we wish to draw your attention to a particular part of a code block, the relevant lines or items are set in bold:

```
func main() {
    blocker := make(chan bool, 1)
    <-blocker
    println("this gets never printed")
}
```

Any command-line input or output is written as follows:

```
tinygo flash –target=arduino-nano33 Chapter06/tinygame/main.go
```

Bold: Indicates a new term, an important word, or words that you see onscreen. For example, words in menus or dialog boxes appear in the text like this. Here is an example: "The value is pretty stable at **37888**."

> **Tips or important notes**
> Appear like this.

Get in touch

Feedback from our readers is always welcome.

General feedback: If you have questions about any aspect of this book, mention the book title in the subject of your message and email us at customercare@packtpub.com.

Errata: Although we have taken every care to ensure the accuracy of our content, mistakes do happen. If you have found a mistake in this book, we would be grateful if you would report this to us. Please visit www.packtpub.com/support/errata, selecting your book, clicking on the Errata Submission Form link, and entering the details.

Piracy: If you come across any illegal copies of our works in any form on the Internet, we would be grateful if you would provide us with the location address or website name. Please contact us at copyright@packt.com with a link to the material.

If you are interested in becoming an author: If there is a topic that you have expertise in and you are interested in either writing or contributing to a book, please visit authors.packtpub.com.

Reviews

Please leave a review. Once you have read and used this book, why not leave a review on the site that you purchased it from? Potential readers can then see and use your unbiased opinion to make purchase decisions, we at Packt can understand what you think about our products, and our authors can see your feedback on their book. Thank you!

For more information about Packt, please visit packt.com.

1
Getting Started with TinyGo

In my opinion, Go is easy to learn, easy to read, and easy to write. The language is not overloaded with fancy features but rather focuses on being concise. The built-in concurrency, fast compile times, high execution performance, and rich standard libraries make a great mix for an awesome language. This is why I want to take you on a journey from very basic high-level Go programs to the depths of microcontrollers utilizing the full power of TinyGo.

In this chapter, we are going to set up TinyGo and learn how to get code completion to work in VS Code and different editors. After this is done, we will have a look at the Arduino UNO and its technical specifications. We are going to compare TinyGo with Go and talk about what makes TinyGo special compared to other languages on microcontrollers. At the end of this chapter, we will write, compile, deploy, and run our first TinyGo program on a real microcontroller. Having all these topics covered, you will have learned how to write, build, and run programs on microcontrollers.

In this chapter, we're going to cover the following main topics:

- Understanding what TinyGo is
- Setting up TinyGo
- Understanding the IDE integration

- The Arduino UNO
- Checking out the Hello World of things

Technical requirements

In order to continue, you need to have the following:

- Go must be installed
- GOPATH must be set up
- Git must be installed
- An Arduino Uno, preferably the Rev3 Edition but you can also use other Arduino Uno boards

You can find all code examples from this chapter in the following GitHub repository: `https://github.com/PacktPublishing/Creative-DIY-Microcontroller-Projects-with-TinyGo-and-WebAssembly/tree/master/Chapter01`

The Code in Action video for the chapter can be found here: `https://bit.ly/3mLFCCJ`

Understanding what TinyGo is

TinyGo is an independently written *compiler*, with its *own runtime implementation*. It is intended to be used for microcontroller programming, **WebAssembly** (**WASM**), and CLI tools. TinyGo heavily makes use of the LLVM infrastructure to optimize and compile code to binaries that a microcontroller can understand.

The first release of TinyGo (v0.1) was published on February 1, 2019 on GitHub. Since then, the project has quickly implemented lots of features and never stopped adding support for more microcontrollers, sensors, displays, and other devices.

On February 2, 2020, TinyGo announced that it is now officially a Google-sponsored project. This was a big step for the complete project.

How TinyGo works

The TinyGo compiler uses a different set of steps than other languages to transform Go source code to machine code. We will not be going into the details though, but let's take a look at an overview of the compiler pipeline:

1. We write the Go source code.
2. This source code gets translated to Go **SSA (Static Single Assignment)**.
3. The Go SSA is transformed to LLVM IR by the TinyGo compiler package.
4. The initialization code in the LLVRM IR is interpreted by the TinyGo `interp` packages. This step optimizes globals, constants, and more.
5. The result is then optimized by some LLVM optimization passes (such as `string` to `[]byte` optimization).
6. The result is then again optimized by the LLVM optimizer.
7. Next, some fixes are done by the compiler package.
8. And as the last step, LLVM creates the machine code.

If this sounds complicated right now, don't worry – we don't have to take care of this process. TinyGo does all this for us. Now let's have a look at what makes TinyGo special compared to Go.

Comparing TinyGo to Go

TinyGo can compile some, but not all Go programs. Let's have a look at an example that can be compiled by both. Let's write a small Hello World program in Go—build it and check its size:

1. This is the most minimal Hello World program I can currently think of:

```
package main
func main() {
    print("Hello World\n")
}
```

It does not need an external package such as `fmt` to print the line.

2. I will be using Go 1.15.2 on an Ubuntu 20.01 operating system. To check your currently installed Go version, use the go version command:

    ```
    $ go version
    go version go1.15.2 linux/amd64
    ```

3. We build the program using the go build command:

    ```
    $ go build ./ch1/hello-world/
    ```

4. Now we check the size using the ls -l command:

    ```
    $ ls -l
    -rwxrwxr-x 1 tobias tobias 1231780 Okt  4 19:31 hello-world
    ```

So, the program has 1,231,780 bytes, which is 1.23178 megabytes. That is pretty big for a program that consists of just 4 lines of code.

> **Note**
>
> The ls command is not available on all operating systems. If you want to check the sizes for yourself, you need to use tools that are available on your operating system.
>
> The size of the binary file may differ when you try it out, as the Go team continues to optimize the compiler.
>
> Furthermore, the size of the binary file could differ when building for other operating systems.

Now let's check what the size of the same program is, but this time compiled using TinyGo. As TinyGo does not support building binaries for Windows, I take care of the compiling, so we can just compare the sizes here:

1. I used the following command to build the binary:

    ```
    $ tinygo build -o hello-world-tiny ch1/hello-world/main.go
    ```

    The tinygo build command has a syntax that is similar to the Go build command.

2. Then I checked the size using the ls -l command, as we did before:

    ```
    $ ls -l
    -rwxrwxr-x 1 tobias tobias 1231780 Okt  4 19:31 hello-
    ```

```
world
-rwxrwxr-x 1 tobias tobias   21152 Okt  4 19:39 hello-
world-tiny
```

We see that the TinyGo version of our Hello World program is only a fraction of the size that the Go compiler emitted. The TinyGo version is only 21,152 bytes, which is about 0.021152 megabytes. The TinyGo program is 58 times smaller as compared to the Go program. This is a huge difference. If you still want to test it out yourself, you can do this after setting up TinyGo.

We have now learned that TinyGo can compile some, but not all Go programs. Also, we learned that programs that are compiled with TinyGo are very small. In the next sections, we'll get to know why TinyGo cannot compile all Go programs and what features TinyGo offers that Go does not offer.

Supported language features

TinyGo supports a part of the Go language features, but not everything is supported right now. Goroutines and channels work on most microcontrollers. Reflection is supported for most types. While slices are supported, you may encounter some problems when working with maps. Only certain types of strings, integers, pointers, and structs or arrays that contain the previous types are supported. So, all in all, a good portion of Go is supported in TinyGo.

Supported standard packages

The biggest part of the standard library is also supported in TinyGo. As of the time of writing, however, most of the `net` and `crypto` packages still do not compile. That means, if you import them, you will get compile errors.

You can look up a list of currently supported standard packages here: `https://tinygo.org/lang-support/stdlib/`.

> **Note**
> A *yes* in the support table does not mean that every function in a package is actually usable in TinyGo. Some functions still could cause compile errors.

Volatile operations

Volatile operations can be used to read to and write from memory-mapped registers. The values inside these registers can change between several reads without the knowledge of the compiler. The compiler has no knowledge about the effects of these operations, hence they are called volatile.

Go does not have a volatile operator, which is why TinyGo provides a **volatile package**. For most cases, we will not need volatile operations, as these are abstracted away by the machine package.

Inline assembly

Assembly Language (**ASM**) is a language that is specifically designed for a certain processor architecture. This happens because assembly depends on the machine code instruction set. The device-specific packages of TinyGo provide assembly packages.

This enables us to use inline assembly code in our Go programs, which is not possible in standard Go.

Heap allocations

The **Heap** is the part of the memory where dynamic allocations and deallocations take place during runtime. So, when our application wants to reserve a part of the memory, it talks with the Heap to reserve the memory. That part of the memory will then be marked as being in use. As this space is rather limited on microcontrollers and garbage collection is expensive and slow, TinyGo tries to optimize away Heap allocations. The result is that, often, objects can be statically allocated instead of dynamically.

Garbage collection

Garbage collection is the process of freeing memory. So, when your application no longer needs a part of the memory it earlier requested, this memory is marked as unused (free) again.

For that purpose, TinyGo has implemented its own variant of garbage collection. TinyGo uses a conservative mark/sweep garbage collection, where conservative means that the **Garbage Collector** (**GC**) has no knowledge of what is a pointer and what it is not. The GC process is split into two parts:

- **Mark**: In the marking phase, the gc marks objects as reachable.
- **Sweep**: In the sweeping phase, the gc frees memory by marking the areas of unreachable objects as free. These freed areas can then be reused to allocate new objects.

We do now know what TinyGo is and what differences exist between TinyGo and Go. We have also learned what the Heap, the GC, and volatile packages are. The next logical step is to go on and set up TinyGo, which we will be doing in the next section.

Setting up TinyGo

The easiest way to install TinyGo and all its dependencies is to follow the Quick Start Guides for Linux, macOS, Windows, and Docker at the following link: `https://tinygo.org/getting-started/`.

As these guides cover important parts, I will only cover the Quick Start part for x64-based architectures and only for Debian-based operating systems such as Ubuntu for Linux.

The first thing to do before we start the setup is to check the latest version of TinyGo. To do so, go to `https://github.com/tinygo-org/tinygo/releases` and check for the newest release. Now, keep this information written down somewhere or memorize it as we'll be using it later.

Installing on Linux

The following steps cover installing TinyGo on a Linux derivate, which is based on Debian:

1. We use the following command to download the `deb` package from GitHub and install it using `dpkg`:

    ```
    wget https://github.com/tinygo-org/tinygo/releases/
    download/v0.15.0/tinygo_0.15.0_amd64.deb
    sudo dpkg -i tinygo_0.15.0_amd64.deb
    ```

    You can exchange the version in the path and filename with the newest release version you found before.

2. Now we must add TinyGo to `GOPATH`. You can use the following command:

    ```
    export PATH=$PATH:/usr/local/tinygo/bin
    ```

    You can also extend `GOPATH` by editing your `.profile` or `.netrc` file.

3. The next step is to verify the installation. Use the `tinygo version` command to verify that TinyGo has been successfully installed:

    ```
    $ tinygo version
    tinygo version 0.15.0 linux/amd64 (using go version
    go1.15.2 and LLVM version 10.0.1)
    ```

4. AVR dependencies: As we are going to work with an Arduino UNO in the first chapters, we need to install some additional dependencies. We do so by using the following commands:

```
sudo apt-get install gcc-avr
sudo apt-get install avr-libc
sudo apt-get install avrdude
```

After installing these dependencies, we can now compile on AVR-based boards such as the Arduino UNO.

If you are using Fedora, Arch Linux, or other distributions, please follow the installation guide: `https://tinygo.org/getting-started/linux/`.

Installing on Windows

In this section, we are going to learn how to install TinyGo on Windows. After this section, we will have also learned how to install dependencies, which are needed to flash the Arduino UNO.

> **Very important note**
>
> You cannot create Windows binary programs using TinyGo. You can still compile and flash programs for microcontroller and WebAssembly targets.
>
> You may want to directly install and use TinyGo inside the **Windows Subsystem for Linux** (**WSL**). The WSL is the way I recommend for Windows users.

To install TinyGo on Windows without using the WSL, I recommend using Scoop, a command-line installer for Windows. Make sure that you have PowerShell 5 (or later) and .NET Framework 4.5 (or later) installed. To do so, please follow these steps:

1. Enable PowerShell for your current user account using the following command:

```
Set-ExecutionPolicy RemoteSigned -scope CurrentUser
```

2. Now run the following command to download Scoop:

```
iwr -useb get.scoop.sh | iex
```

3. You can install TinyGo by using the following command:

```
scoop install tinygo
```

4. Now to verify the installation was successful, use the following command:

```
tinygo version
```

The output should look like the following:

```
tinygo version 0.15.0 windows/amd64 (using go version
go1.15.3 and LLVM version 10.0.1)
```

The actual TinyGo and Go version might differ.

5. AVR dependencies: In order to be able to compile and flash programs for the Arduino UNO, we need to have an AVR 8-bit toolchain installed. You can find a download here: https://www.microchip.com/mplab/avr-support/avr-and-arm-toolchains-c-compilers.

Extend your %PATH% and make sure that the bin folder is included:

6. Next, download and install GNU Make for Windows. You can find GNU Make here: http://gnuwin32.sourceforge.net/packages/make.htm.

7. As the last step, you need to download and install avrdude. The avrdude EXE also must be inside your %PATH%. You can download AVR Dude here: http://download.savannah.gnu.org/releases/avrdude/. The file you are looking for is called avrdude-6.3-mingw32.zip.

If you encounter any problems regarding the avr setup or don't know how to configure environment variables, you may want to check out the following guide: https://fab.cba.mit.edu/classes/863.16/doc/projects/ftsmin/windows_avr.html.

> **WSL installation**
>
> It is also possible to install TinyGo directly on the **Windows Subsystem for Linux (WSL)**. Just follow the Linux section to do that.

Installing on macOS

Installation on macOS is straightforward. Let's take a quick look at the steps:

1. We are going to use Homebrew to install tinygo. Just use the following two commands:

```
brew tap tinygo-org/tools
brew install tinygo
```

2. Simply run the `tinygo version` command to verify the installation:

```
$ tinygo version
tinygo version 0.15.0 darwin/amd64 (using go version
go1.15 and LLVM version 10.0.1)
```

3. Run the following commands to install the additional requirements needed to compile programs for AVR-based microcontrollers such as the Arduino UNO, which we are going to use in the first few chapters of the book:

```
brew tap osx-cross/avr
brew install avr-gcc
brew install avrdude
```

Installing on Docker

It is possible to directly use a Docker image to compile our programs. However, it is not possible to flash the programs using the Docker image.

Simply download the image using the following:

```
docker pull tinygo/tinygo:0.15.0
```

> **Note**
>
> The actual TinyGo version might differ. Use the newest TinyGo version from the check we did when we started the section.

Here is an example call to build a program:

```
docker run -v $GOPATH:/go -e "GOPATH=/go" tinygo/tinygo:0.15.0
tinygo build -o /go/src/github.com/myuser/myrepo/wasm.wasm
-target wasm --no-debug /go/src/github.com/myuser/myrepo/wasm-
main.go
```

We have now successfully set up TinyGo and installed all additional requirements to compile and flash programs to the Arduino UNO microcontroller. Also, everything we need for WebAssembly is now set up. The next step is to set up IDE integration before we start writing our first program for a microcontroller.

Understanding IDE integration

Having a properly set up IDE is truly a blessing as we benefit from its features of code completion, functional linting, and so on. This way, we do not have to investigate the source code or documentation for every function we want to call.

In this section, we will look at the process of integrating TinyGo into VS Code, Goland, and other editors. This enables us to choose whatever editor we prefer to use.

VS Code integration

VS Code offers an extension system, which makes it easy to integrate the Go and TinyGo toolset into the IDE. We are going to install the Go Extension, which offers support for the Go programming language. Afterward, we are going to install the TinyGo extension, which brings TinyGo support.

The Go extension

We install the Go extension using the **Extensions** view using the following steps:

1. Open the **Extensions** view either by clicking on the **Extensions** icon or pressing *Ctrl + Shift + X*.

2. Search for **Go**.

3. Select the first entry in the list, which is called **Go** and is from the Go team at Google.

4. Click on the **Install** button, as seen in the following screenshot:

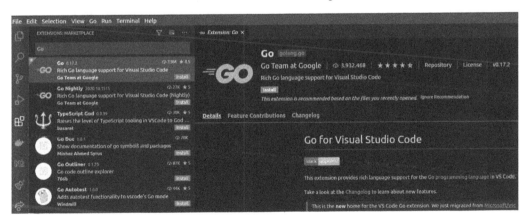

Figure 1.1 – Installation from the Extensions view

5. After installing the extension for the first time, you might get prompted to install more dependencies. Do so by clicking on the **Install** button. If you do not get prompt, you can also install all dependencies by hitting *Ctrl + Shift + P* and entering the following command:

```
Go: install
```

6.   Select the **Go: Install/Update Tools** command and hit *Enter*:

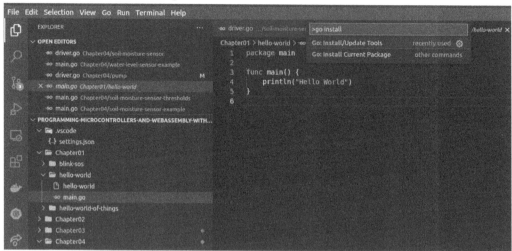

Figure 1.2 – Go: Install/Update Tools command to execute

7.   Now select all dependencies by checking the box on the left side and click **OK**:

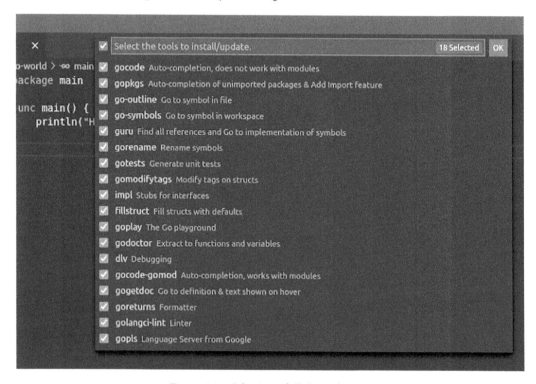

Figure 1.3 – Selection of all dependencies

8. VS Code will now install all dependencies, and it should print the following message when done:

```
All tools successfully installed. You are ready to Go :).
```

Next, we'll see the TinyGo integration in VS Code.

The TinyGo extension

TinyGo integration in VS Code is straightforward as there is a TinyGo extension that we simply need to install. Let's quickly go through the steps:

1. Open the **Extensions** view either by clicking on the **Extensions** icon or pressing *Ctrl + Shift + X*.

2. Search for **TinyGo**.

3. Select the first entry in the list, which is called **TinyGo** and is from the TinyGo team.

4. Click on the **Install** button, as seen in the following screenshot:

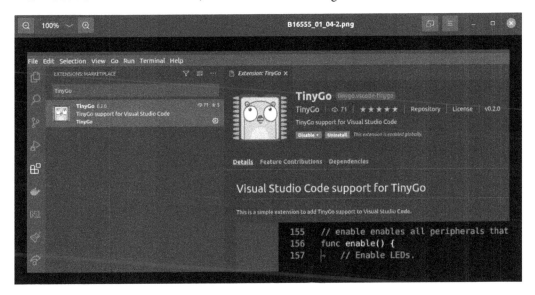

Figure 1.4 – Extensions view showing the TinyGo extension

5. We are not done with installing the extension. We need to use another command to configure the target we want to build for. Hit *Ctrl + Shift + P*, type TinyGo target, and hit *Enter*.

6. Now search for `arduino` and hit *Enter*, as we see in the following screenshot:

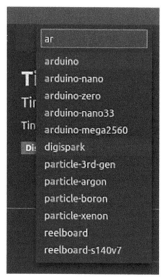

Figure 1.5 – The target selection popup

7. VS Code will open a popup telling you that it needs to reload the window. Do so by clicking on **Reload**:

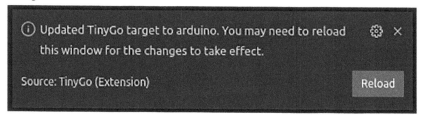

Figure 1.6 – Popup asking to reload the window

Okay, we now have the extension installed and have selected a target. But what does it do internally? The only function of this extension is to set the `go.toolsEnvVars` variable in `vs code settings.json` of your current project.

This could look like the following example:

```
{
    "go.toolsEnvVars": {
        "GOROOT": "/home/user/.cache/tinygo/goroot-go1.14-f930d5b
            5f36579e8cbf1f-syscall",
        "GOFLAGS": "-tags=cortexm,baremetal,linux,arm,nrf51822,
            nrf51,nrf,microbit,tinygo,gc.conservative,scheduler.
            tasks"
```

```
    }
  }
```

Sometimes a popup similar to the one in the following screenshot will appear. Do not click on **Update tools**; just close it.

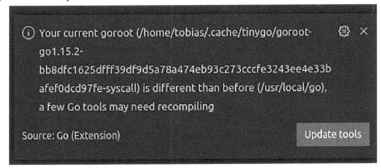

Figure 1.7 – Popup asking to update tools

If you are using VS Code, congratulations, you are done with the setup and are ready to go! The next sections are going to explain how to set up the IDE integration in other editors.

General IDE integration

You may wonder, how does IDE integration work with TinyGo? Well, we simply have to configure the standard Go tooling, especially the **gopls language** server.

TinyGo has its own implementation of the standard libraries and also provides additional libraries, such as the machine package. The gopls language server needs to know where to look for these packages. That is why we need to set a GOROOT for this project.

TinyGo makes heavy use of compiler flags. These flags are used during compile time to determine which files must be included in the build, as we see in the following screenshot:

```
src > machine > ∞ board_arduino.go > ▤ D0
  1    // +build arduino
  2
  3    package machine
  4
  5    // Return the current CPU frequency in hertz.
  6    func CPUFrequency() uint32 {
  7        return 16000000
  8    }
  9
```

Figure 1.8 – The board_arduino.go file from the TinyGo source code showing a build flag

So basically, we integrate TinyGo into an IDE by locally setting these environment variables.

We do not have to guess the correct values for GOROOT and GOFLAGS. TinyGo provides a command for that purpose. Let's say we want to set the correct flags for an Arduino, we can find out by using the following command:

```
tinygo info arduino
```

This will print the following result:

```
LLVM triple:        avr-unknown-unknown
GOOS:               linux
GOARCH:             arm
build tags:         avr baremetal linux arm atmega328p atmega
avr5 arduino tinygo gc.conservative scheduler.none
garbage collector:  conservative
scheduler:          none
cached GOROOT:      /home/tobias/.cache/tinygo/goroot-go1.15.2-
bb8dfc1625dfff39df9d5a78a474eb93c273cccfe3243ee4e33bafef0dcd9
7fe-syscall
```

The important parts of the output are build tags and cached GOROOT.

As we now know where to find the needed information, we can go ahead and configure any IDE we want to use.

Setting up Goland

As we have now learned that we must set a GOROOT and build tags, we can also configure the integration in Goland.

Set **GOROOT** to the cached GOROOT from the tinygo info command, as seen in the following screenshot:

Figure 1.9 – GOROOT configuration in Goland

The next step is to set the `build` tags. You can find them under **Build Tags & Vendoring**. Add the tags into the **Custom tags** field:

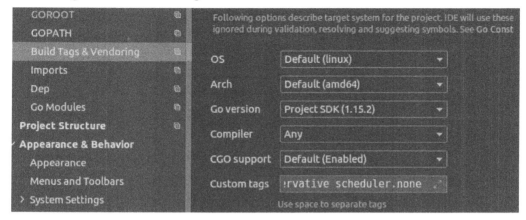

Figure 1.10 – Custom tags configuration in Goland

> **Note**
>
> You must manually change **Custom tags** each time you want to program for another microcontroller.

Integrating any editor

If you have the standard Go tolling installed, you can use any other editor such as Vim or Nano so you can get IDE support. As other editors might lack configuration files, we can work around this fact by passing them the environment variables at the moment we start them.

In the following example, we start a VS Code instance by first setting environment variables and then starting VS Code:

```
export GOFLAGS=-tags=avr,baremetal,linux,arm,atmega328p,atmega
,avr5,arduino,tinygo,gc.conservative,scheduler.none; code
```

You can exchange the code call to any other program such as `vim` or `nano`. On Windows systems, the call might look a little bit different.

As we now know how to configure any IDE for the use of TinyGo, we will move on to learn about the Arduino UNO.

The Arduino UNO

The Arduino UNO is one of the most popular boards out there. It is powered by an 8-bit ATmega328P microcontroller, and as of the writing of this book, there are lots of derivates from the original Arduino UNO boards. Let's get to know it better in the following subsections.

Getting to know the technical specifications

As you can see in the following table, the ATmega328P has only 16 MHz and 32 KB Flash memory. Standard Go produces a Hello World program of about 1.2 MB, which would not even fit on this microcontroller. So, we are working with very limited hardware here, but you will see that this is sufficient to build amazing projects.

Here is a brief look at the technical specs of the Arduino UNO:

Operating voltage	5V
Input voltage limits	6-20V
Recommended input voltages	7-12V
Digital I/O pins	14 (with 6 of them with PWM output)
Analog input pins	6
DC current per I/O pin	20 mA
DC current for 3.3V pin	50 mA
Flash memory	32 KB
SRAM	2 KB
Clock speed	16 MHz
Built-in LED pin	13

Table 1.1 – Technical specifications

> **Note**
> Consider the DC current per I/O pin of 20 mA as an upper limit. You should not exceed this limit to prevent damaging your microcontroller.

Let's have a look at the pinout next.

Exploring pinout

A **Pinout** is basically a map of the pins. We are going to use the descriptions of these pins in all projects that we build with the Arduino UNO. We will need it to correctly wire our components.

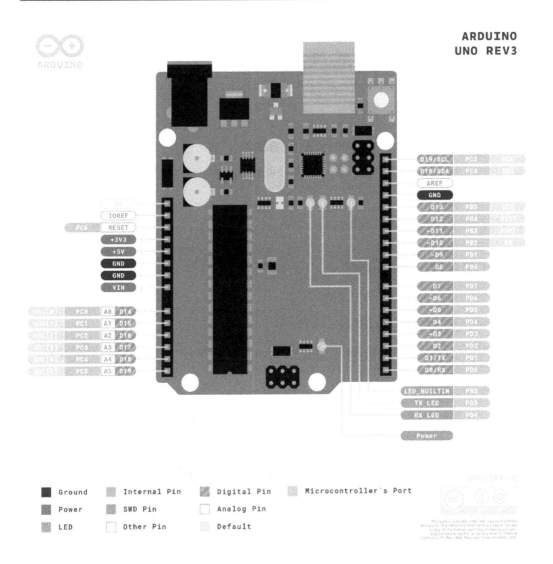

Figure 1.11 – Arduino UNO REV3 pinout

As we now have learned some basic information about the Arduino UNO, let's go on and write our first program.

Checking out the Hello World of Things

A Hello World program is the typical way to start the journey in a new programming language. A Hello World program on a microcontroller looks a bit different compared to a normal Hello World program. We are going to write a Hello World program to let the built-in LED blink. Let's get started!

Getting the requirements ready

To get started with our program, we need the following:

- An Arduino UNO
- One USB cable to connect it to your computer

Preparing the project

Follow these steps closely for your project:

1. Create a new folder named ch1 in the root of your project.
2. Inside the folder, we need to create a folder named hello-world-of-things and inside it, we are going to create a new main.go file.
3. Your structure should now look like the following:

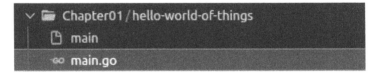

Figure 1.12 – The project structure

As we have now prepared our project, we can go on and write our first program.

Programming the microcontroller

We are going to let the onboard LED blink. It is the easiest possible start for us. The example we are using is inspired by the *Blinky example* from the TinyGo source code, which is also used as a Hello World of Things showcase. Let's go through each step carefully:

1. Declare the main package:

    ```
    package main
    ```

2. Import the packages `machine` and `time`:

```
import (
    "machine"
    "time"
)
```

3. Add a `main` function:

```
func main() {
```

4. Initialize a variable named `led` with the value `machine.LED`:

```
led := machine.LED
```

5. Configure the `led` pin as the output pin:

```
led.Configure(machine.PinConfig{Mode: machine.
PinOutput})
```

6. Declare an endless loop:

```
for {
```

7. Set the `led` to `Low` so that no voltage is given on the LED:

```
led.Low()
```

8. Set `sleep` for `300` milliseconds, that is the time the LED is off:

```
time.Sleep(time.Millisecond * 300)
```

9. Set the `led` to `High` so a voltage is given to the LED to make it shine:

```
led.High()
```

10. Set `Sleep` for `300` milliseconds, which is the amount of time the LED is on for:

```
time.Sleep(time.Millisecond * 300)
```

11. Close the `for` loop:

```
}
```

12. Closing braces for the `main` function:

```
}
```

The `machine` package provides constants for the pin mapping and provides some more functions that are directly related to the used microcontroller.

We must wait for a certain amount of time between giving voltage to the LED and taking it off again, so we can see the blinking.

Configuring a pin as output means that we tell the microcontroller that we are only going to send signals using this pin. We can also configure a pin as input, which enables us to read the state from the pin.

Flashing the program

Flashing the program is a simple command, if you are on Linux, macOS, or are using Windows WSL.

Simply connect your Arduino UNO to any USB port and execute the following command:

```
tinygo flash --target=arduino ch1/hello-world-of-things/main.go
```

The `tinygo flash` command needs at least the following parameters:

- `--target`, which sets the microcontroller to flash
- The path to the `main.go` file

Your output should look like the following:

```
avrdude: AVR device initialized and ready to accept
instructions
Reading | ############################################ |
100% 0.00s
avrdude: Device signature = 0x1e950f (probably m328p)
avrdude: NOTE: "flash" memory has been specified, an erase
cycle will be performed
         To disable this feature, specify the -D option.
avrdude: erasing chip
avrdude: reading input file "/tmp/tinygo208327574/main.hex"
avrdude: writing flash (558 bytes):
Writing | ############################################ |
100% 0.10s
avrdude: 558 bytes of flash written
avrdude: verifying flash memory against /tmp/tinygo208327574/
main.hex:
```

```
avrdude: load data flash data from input file /tmp/
tinygo208327574/main.hex:
avrdude: input file /tmp/tinygo208327574/main.hex contains 558
bytes
avrdude: reading on-chip flash data:
Reading | ############################################### |
100% 0.08s
avrdude: verifying ...
avrdude: 558 bytes of flash verified
avrdude done.   Thank you.
```

As you can see, in my example, the code flashed onto the Arduino UNO is only using 558 bytes of memory.

Congratulations, you have successfully written, built, and flashed your first program onto an Arduino UNO using TinyGo.

Using TinyGo Playground

You don't have an Arduino UNO right now? You can test the code using **TinyGo Playground**. TinyGo Playground makes use of WebAssembly to emulate the behavior of a small number of boards such as the Arduino Nano IoT 33 and the Arduino UNO. It can also compile programs for the Arduino Nano IoT 33. But please keep in mind that the behavior in the TinyGo Playground might differ from real hardware.

You can find the TinyGo Playground at https://play.tinygo.org/.

Summary

We have learned what TinyGo actually is, how it differs from standard Go, we have acquired basic knowledge about the Arduino UNO itself, how to set up TinyGo, how to set up IDE integration, and finally, wrote and flashed our first program onto real hardware and made an LED blink with our code. Isn't that an interesting start?

We are going to build a traffic light controller system in the next chapter.

Questions

1. Which command can be used to find out the needed environment variable values for the IDE integration?

2. Which command can be used to flash a program onto a microcontroller?

3. Why do we have to sleep a certain amount of time when giving voltage or taking voltage away from the LED?

4. How would you let the LED blink S-O-S in morse code?

2
Building a Traffic Lights Control System

In the previous chapter, we set up TinyGo and our IDE, and we now know how to build and flash our programs to the Arduino UNO. We are now going to utilize this knowledge to go one step further.

In this chapter, we are going to build a traffic lights control system. We are going to split the project into small steps, where we build and test each component. At the end, we are going to put everything together. We will be using multiple LEDs, a breadboard, GPIO ports, and a button to interrupt the normal flow to switch pedestrian lights to green. By the end of the chapter, you will know how to control external LEDs, read the state of a button, use GPIO ports, how to distinguish resistors, and how to utilize Goroutines in TinyGo.

In this chapter, we are going to cover the following topics:

- Lighting an external LED
- Lighting a single LED when a button is pressed

- Building traffic lights
- Building traffic lights with pedestrian lights

Technical requirements

To build the traffic lights control system, we are going to need some components. We will need the following to build the complete project:

1. An Arduino UNO
2. Breadboard
3. Five LEDs
4. Multiple jumper cables
5. Multiple 220 Ohm resistors
6. One push button
7. One 10K Ohm resistor

You can find all code examples from this chapter in the following GitHub repository: `https://github.com/PacktPublishing/Creative-DIY-Microcontroller-Projects-with-TinyGo-and-WebAssembly/tree/master/Chapter02`

The Code in Action video for the chapter can be found here: `https://bit.ly/2RpvF2a`

Lighting an external LED

Before we start to build a more complex circuit, let's begin with lighting up an external LED. As soon as this is working, we are going to extend the circuit step by step. We begin with a single red LED. Lighting up an external LED is a bit different compared to lighting up an onboard LED. We are going to need something on which we can place the LED, and we will need some wires as well as a basic understanding of resistors, which will help us to prevent the LED from taking damage. That is why we are going to look at each component one by one.

Using breadboards

Breadboards are used for prototyping, as they do not require you to directly solder components. We are going to build all our projects using breadboards.

A breadboard typically consists of two parts:

- The power bus
- Horizontal rows

Each side of a breadboard has a power bus. The power bus provides a + (positive) lane and a - (ground) lane. The positive lane is colored *red* and the ground lane is colored *blue*. The individual slots are connected inside the power bus.

The slots of a single horizontal row are also connected. A signal in one slot is also available in the next slot. Different horizontal rows are not connected, unless we put a cable in there to create a connection. Here's what a breadboard looks like:

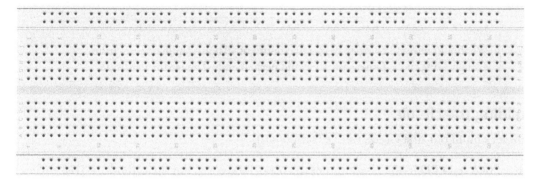

Figure 2.1 – A breadboard – image taken from Fritzing

Understanding LED basics

The Arduino UNO has an operating voltage of 5V, which is too high for most LEDs. So, we need to reduce the voltage to something our LEDs can handle. For that reason, we will be using 220 Ohm resistors to draw current from the line in order to protect the LED from damage. If you do not have 220 Ohm resistors, you can also use 470 Ohm as well; anything between 220 and 1K (1K = 1,000) Ohm will be fine.

If you want to really make sure that the resistor matches the needs of the LED, you can also calculate the resistor value as follows:

$R = (V_s - V_{led}) / I_{led}$

Where:

- R is the resistor value.
- V_s is the source voltage.
- V_{led} is the voltage drop across the LED.
- I_{led} is the current through the LED.

> **Note**
>
> LEDs have *anode* (+) and *cathode* (-) leads. The anode lead is longer.
>
> Different colors need to be driven with different voltages. When using the same resistors and voltages for the different LED colors, you will notice that some colors will be brighter compared to others.

Using GPIO ports

GPIO stands for **General Purpose Input Output**. That means we can use these pins for input as well as output for digital signals. We can either set a GPIO pin to *High* or *Low*, or read a *High* or *Low* value from the port.

> **Note**
>
> We should never draw more than a maximum of 40.0 mA (milliampere) from a single GPIO port. Otherwise, we could permanently damage the hardware.

Building the circuit

Now let's build our first circuit on the breadboard:

1. Put a red LED in the *G* column of the horizontal rows. Put the cathode in *G12* and the anode in *G13*.

2. Connect *F12* with the ground lane on the power bus.

3. Connect *F13* and *E13* using a 220 Ohm resistor. (Anything between 220 and 1,000 Ohms is okay.)

4. Connect *Pin 13* from the GPIO ports to *A13*.

5. Connect the GND port to the ground lane on the power bus.

> **Note**
>
> The descriptions on your breadboard might differ from the ones I am using. If that is the case, you'll need to build the circuit by checking the next figure.

The circuit should now look like the following:

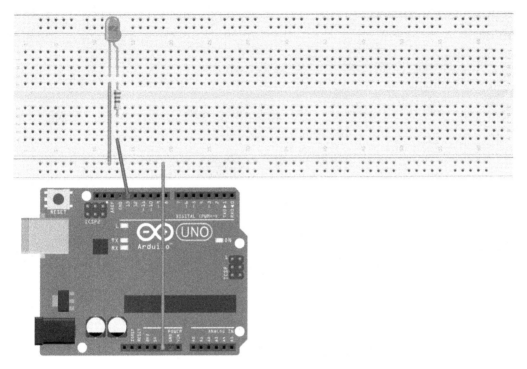

Figure 2.2 – Image of the circuit – image taken from Fritzing

Writing the code

We start off by creating a new folder named Chapter02 in our project workspace. This folder will be used for all parts of this chapter. Inside the Chapter02 folder, we create a blinky-external folder and create a new main.go file inside.

The structure should look like the following:

Figure 2.3 - Project structure for writing the code

We import the `machine` and `time` packages and put the following code into the `main` function:

1. Declare and initialize a variable named `outputConfig` with a new `PinConfig` in output mode:

   ```
   outputConfig := machine.PinConfig{Mode: machine.
                   PinOutput}
   ```

2. Declare and initialize a variable named `greenLED` with a value of `machine.D13`:

   ```
   greenLED := machine.D13
   ```

3. Configure the LED with the `outputConfig` instance we created earlier, by passing it as a parameter into the `Configure` function:

   ```
   redLED.Configure(outputConfig)
   ```

4. We then loop endlessly:

   ```
   for {
   ```

5. Set `redLED` to Low (off):

   ```
   redLED.Low()
   ```

6. Sleep for half a second. Without sleeping, the LED would be turned on and off at an extremely high rate, so we sleep after each change in a state:

   ```
   time.Sleep(500 * time.Millisecond)
   ```

7. Set the `redLED` to High (on):

   ```
   redLED.High()
   ```

8. Sleep for half a second:

   ```
   time.Sleep(500 * time.Millisecond)
   }
   ```

9. Now flash the program using the `tinygo flash` command using the following command:

   ```
   tinygo flash –target=arduino Chapter02/blinky-external/
   main.go
   ```

When the flash progress completes and the Arduino restarts, the red LED should blink at intervals of 500 ms.

Congratulations, you have just built your first circuit and written your first program to control external hardware! As we now know how to connect and control external LEDs on a breadboard, we can continue to build a more advanced circuit. Let's do just that in the next section.

Lighting an LED when a button is pressed

Until now, we have only used code to directly control hardware components. Let's now try to read the state of a button in order to control an LED. We will need the following components:

- At least 6 jumper wires
- One LED (the color does not matter)
- One 220 Ohm resistor
- One 4-pinned-button (push down button)
- One 10K Ohm resistor

Now let's go on to build the circuit.

Building the circuit

The following circuit extends the one we previously built. So, if you still have the previous circuit assembled, you just have to add the button part. The next circuit consists of two component groups. The first group is used to control an LED, and the second group is used to read the button state.

Adding the LED component

We start off with the LED circuit:

1. Place an LED with the cathode in G12 and the anode in G13.
2. Use a 220 Ohm resistor to connect *F13* with *D13*.
3. Connect port *D13* from the GPIO ports with *A13* using a jumper wire.
4. Connect *F12* with the ground lane of the power bus using a jumper wire.

Adding the button component

Now we are going to add a button:

1. Use a jumper wire to connect A31 with the positive lane of the power bus.

2. Use a 10K Ohm resistor to connect the ground lane of the power bus with B29.

3. Connect D29 with port D2.

4. Place the push button with one pin in E29, one in E31, one in F29, and the last pin in F31.

Our circuit should now look similar to the following:

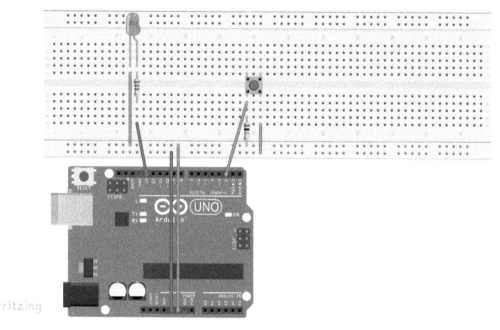

Figure 2.4 – The circuit – image taken from Fritzing

Note

Before we start to write the code for this circuit, we need to learn how these buttons work.

As the button will not work if you place it incorrectly onto the breadboard, let's have a look at the button again.

The 4 pins on the button are grouped into two pins each. So, two pins are connected to each other. Looking at the back of the button, we should be able to see that two opposing pins are connected to each other. So, the button won't work as expected when you place it rotated by 90°.

Programming the logic

Before diving into the code, we will create a new folder named `light-button` inside the `Chapter02` folder and create a `main.go` file in it, with an empty `main` function, using the following:

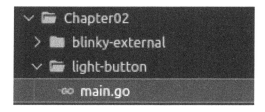

Figure 2.5 – The folder structure for the logic

Let's now look at the `main` function and the pull-up resistor.

The main function

We want to light the LED when the button is pressed. To achieve this, we need to read from a pin and check for its state using the following steps:

1. Initialize the `outPutConfig` variable with `PinConfig` in `PinOutput` mode. This config is going to be used to control the LED pin:

    ```
    outputConfig := machine.PinConfig{Mode: machine.
                    PinOutput}
    ```

2. Initialize the `inputConfig` variable with `PinConfig` in `PinInput` mode. This config is being used for the pin that reads the button state and therefore needs to be an input:

    ```
    inputConfig := machine.PinConfig{Mode: machine.PinInput}
    ```

3. Initialize the `led` variable with a value of `machine.D13`, which is the pin we have connected to `led`:

    ```
    led := machine.D13
    ```

4. Configure `led` as output by passing `outputConfig` as the parameter, which is the pin that is connected to the button:

    ```
    led.Configure(outputConfig)
    ```

5. Initialize the `buttonInput` variable with a value of `machine.D2`:

    ```
    buttonInput := machine.D2
    ```

6. Configure `buttonInput` as an input by passing `inputConfig` as the parameter:

    ```
    buttonInput.Configure(inputConfig)
    ```

7. As we do not want the program to be terminated after checking the button state a single time, we use an endless loop to repeat and check forever:

    ```
    for {
    ```

8. Check the current state of the button. It will be true if the button is pressed:

    ```
    if buttonInput.Get() {
    ```

9. If the button is pressed, we light up the LED:

    ```
    led.High()
    ```

10. We are calling `continue` here, so we do not execute the `led.Low()` call:

    ```
    continue
    }
    ```

11. If the button is not pressed, we turn the LED off:

    ```
    led.Low()
    }
    ```

> **Note**
>
> Do not forget to import the `machine` package, otherwise the code will not compile.

Now flash the program using the `tinygo flash` command:

```
tinygo flash -target=arduino Chapter02/light-button/main.go
```

After successfully flashing, the LED should light up when you press the button.

The pull-up resistor

You may have wondered why we need a 10K Ohm resistor in the button circuit. The 10K Ohm resistor is used to prevent the signal/pin from floating. Floating pins are bad, as an input pin in a floating state is indeterminate. When trying to read a value from a pin, we expect to get a digital value – 1 or 0, or true or false. Floating means that the value can change rapidly between 1 and 0, which happens without pull-up or pull-down resistors. Here's some further reading on floating pins: `https://www.mouser.com/blog/dont-leave-your-pins-floating`.

As an alternative to the 10K Ohm external resistor, an internal resistor can be used.

Configuring an input pin to use an internal resistor is done as follows:

```
inputConfig := machine.PinConfig{
            Mode: machine.PinInputPullup
}
```

We have now learned how to control an LED using an input signal, which was given by a button. The next step is to build the traffic lights flow to control three LEDs.

Building traffic lights

We know how to light up a single LED, and we also know how to light up an LED using a button input. The next step is to build a circuit using three LEDs and to write the code to light them up in the correct order.

Building the circuit

To build the circuit, we need the following components:

- Three LEDs (preferably red, yellow, and green)
- Three 220 Ohm resistors
- Seven jumper wires

We start by first setting up the components using the following steps:

1. Connect *GND* from the Arduino to any ground port on the power bus.
2. Place the first (red) LED with the cathode in *G12* and the anode in *G13*.
3. Place the second (yellow) LED with the cathode in *G15* and the anode in *G16*.

4. Place the third (green) LED with the cathode in *G18* and the anode in *G19*.

5. Connect *F13* with *D13* using a 220 Ohm resistor.

6. Connect *F16* with *D16* using a 220 Ohm resistor.

7. Connect *F19* with *D19* using a 220 Ohm resistor.

8. Connect *F13* to *Ground* on the power bus using a jumper wire.

9. Connect *F16* to *Ground* on the power bus using a jumper wire.

10. Connect *F19* to *Ground* on the power bus using a jumper wire.

11. Connect port *D13* to *A12* using a jumper wire.

12. Connect port *D16* to *A12* using a jumper wire.

13. Connect port *D19* to *A12* using a jumper wire.

Your circuit should now look similar to the following figure:

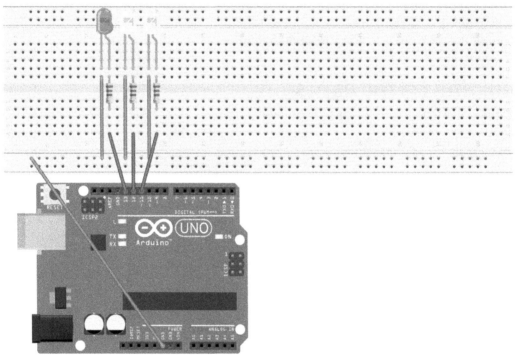

Figure 2.6 – The traffic lights circuit – image taken from Fritzing

We have now successfully set up the circuit. Now we can continue to write some code to control the LEDs.

Creating a folder structure

We start off by creating a new folder named `traffic-lights-simple` inside the `Chapter02` folder. Also, we create a `main.go` file inside the new folder and start off with an empty `main` function. Your project structure should now look like this:

Figure 2.7 - Folder structure for the circuit

Writing the logic

We have successfully set up our project structure to continue. We are going to implement the following flow:

RED -> RED-YELLOW -> GREEN -> YELLOW -> RED

This is a typical flow for traffic lights with three bulbs.

We are going to configure three pins as output, and afterward, we want to endlessly loop and light up the LEDs in this flow.

Inside the `main` function, we write the following:

1. Initialize a new variable named `outputConfig` as `PinConfig` using the `PinOutPut` mode:

    ```
    outputConfig := machine.PinConfig{Mode: machine.
                    PinOutput}
    ```

2. Initialize a new variable named `redLED` with the value `machine.D13` and configure it as output:

    ```
    redLED := machine.D13
    redLED.Configure(outputConfig)
    ```

3. Initialize a new variable named `yellowLED` with the value `machine.D12` and configure it as output:

    ```
    yellowLED := machine.D12
    yellowLED.Configure(outputConfig)
    ```

4. Initialize a new variable named `greenLED` with the value `machine.D11` and configure it as output:

```
greenLED := machine.D11
greenLED.Configure(outputConfig)
```

We have now initialized our variables to act as output pins. The next step is to light up the LEDs in the correct order. We basically have four phases, which just need to repeat in order to simulate a real traffic light. Let's go through these one by one:

1. We are going to handle the phases in an endless loop:

```
for {
```

2. For *RED-Phase*, turn on the red LED and wait for a second:

```
redLED.High()
time.Sleep(time.Second)
```

3. For *RED-YELLOW-Phase*, turn on the yellow LED and wait for a second:

```
yellowLED.High()
time.Sleep(time.Second)
```

4. For *GREEN-PHASE*, turn off the yellow and red LEDs and turn on the green LED and wait for a second:

```
redLED.Low()
yellowLED.Low()
greenLED.High()
time.Sleep(time.Second)
```

5. For *YELLOW-Phase*, turn off the green LED and turn on the yellow LED, then wait for a second and turn off yellow again, so we can start cleanly with *RED-Phase* again:

```
greenLED.Low()
yellowLED.High()
time.Sleep(time.Second)
yellowLED.Low()
}
```

The complete content of the function is available at the following URL:

```
https://github.com/PacktPublishing/Programming-
Microcontrollers-and-WebAssembly-with-TinyGo/blob/master/
Chapter02/traffic-lights-simple/main.go
```

> **Note**
> Don't forget to import the `time` and `machine` packages.

We have now assembled and programmed a complete traffic lights flow. The next step is to combine everything we have built to complete our project.

Building traffic lights with pedestrian lights

We will now combine everything we have learned and done in this chapter to create an even more realistic traffic lights system. We will do so by assembling a circuit that contains the three-bulb traffic lights from the previous step and adding pedestrian lights with two bulbs that are controlled by a button.

Assembling the circuit

For our final project in this chapter, we need the following:

- Five LEDs: preferably two red, one yellow, and two green
- Five 220 Ohm resistors, one for each LED
- One 10K Ohm resistor as a pull-up resistor for our push button
- One 4-pin push button
- 14 jumper wires

We start by setting up the three-bulb traffic lights using the following steps:

1. Place the first LED (red) with the cathode on *G12* and the anode on *G13*.
2. Place the second LED (yellow) with the cathode on *G15* and the anode on *G16*.
3. Place the third LED (green) with the cathode on *G18* and the anode on *G19*.
4. Use a 220 Ohm resistor to connect *F13* with *D13*.
5. Use a 220 Ohm resistor to connect *F16* with *D16*.
6. Use a 220 Ohm resistor to connect *F19* with *D19*.

7. Connect pin *D13* with *A13* using a jumper wire.

8. Connect pin *D12* with *A16* using a jumper wire.

9. Connect pin *D11* with *A10* using a jumper wire.

10. Connect *F12* with Ground on the power bus using a jumper wire.

11. Connect *F15* with Ground on the power bus using a jumper wire.

12. Connect *F18* with Ground on the power bus using a jumper wire.

Now assemble the pedestrian lights using the following steps:

1. Place the fourth LED (red) with the cathode on *G22* and the anode on *G23*.

2. Place the fifth LED (green) with the cathode on *G25* and the anode on *G26*.

3. Use a 220 Ohm resistor to connect *F23* with *D23*.

4. Use a 220 Ohm resistor to connect *F26* with *D26*.

5. Connect pin *D5* with *A23* using a jumper wire.

6. Connect pin *D4* with *A26* using a jumper wire.

7. Connect *F22* with Ground on the power bus using a jumper wire.

8. Connect *F24* with Ground on the power bus using a jumper wire.

Now we only need to assemble the button and connect the power bus:

1. Place a push button with the left pins in *E29* and *F29* and the right pins on *E31* and *F31*.

2. Use a 10K Ohm resistor to connect the Ground from the power bus with *B29*.

3. Connect pin *D2* with *C29* using a jumper wire.

4. Connect *A31* with the positive lane on the power bus using a jumper wire.

5. Connect the positive lane on the power bus with the 5V port on the Arduino UNO using a jumper wire.

6. Connect the ground lane on the power bus with a ground port on the Arduino UNO using a jumper wire.

When you've finished assembling, your circuit should look like this:

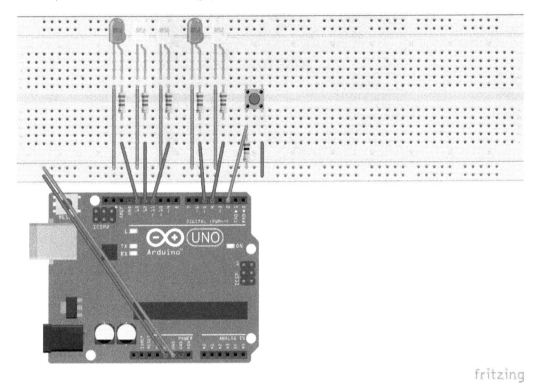

Figure 2.8 – Circuit for the traffic lights with pedestrian lights controlled by
a button – image taken from Fritzing

Great, we have now completely assembled our final project for this chapter. We can now write some code to bring this project to life.

Setting up the project structure

We start off by creating a new folder named `traffic-lights-pedestrian` inside the `Chapter02` folder. Inside the new folder, we create a new `main.go` file with an empty `main` function.

Our project structure should now look like the following:

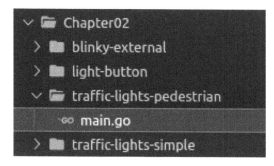

Figure 2.9 - Project structure for the project

Writing the logic

We are going to split the program into three parts:

- Initialization logic
- Main logic
- trafficLights logic

Initializing the logic

We need to initialize a stopTraffic variable and configure the pins for the LEDs as output pins using the following steps:

1. We start off by declaring a bool variable named stopTraffic at the package level. This variable is going to be used as a communication channel between our two logic parts:

   ```
   var stopTraffic bool
   ```

2. The first thing we do in the main method is set the value of stopTraffic to false:

   ```
   stopTraffic = false
   ```

3. We declare and initialize a new variable named outputConfig with PinConfig in PinOutput mode. We are going to pass this config to all LED pins:

   ```
   outputConfig := machine.PinConfig{Mode: machine.
                    PinOutput}
   ```

4. We initialize some new variables: greenLED with the value machine.D11, yellowLED with the value machine.D12, and redLED with the value machine.D13. Then, we configure each LED variable as output pins:

```
greenLED := machine.D11
greenLED.Configure(outputConfig)
yellowLED := machine.D12
yellowLED.Configure(outputConfig)
redLED := machine.D13
redLED.Configure(outputConfig)
```

5. We initialize some new variables: pedestrianGreen with the value machine.D4 and pedestrianRed with the value machine.D5. Then, we configure each LED variable as output pins:

```
pedestrianGreen := machine.D4
pedestrianGreen.Configure(outputConfig)
pedestrianRed := machine.D5
pedestrianRed.Configure(outputConfig)
```

6. We declare and initialize a new variable named inputConfig with PinConfig in PinInput mode. Then, we declare and initialize a new variable named buttonInput with the value machine.D2 and configure buttonInput as the input pin:

```
inputConfig := machine.PinConfig{Mode: machine.PinInput}
buttonInput := machine.D2
buttonInput.Configure(inputConfig)
```

That's it for the initialization. We have set up all pins and a Boolean variable at the package level.

> **Note**
> The pin constants, such as machine.D13, are of the machine.Pin type.

Writing the trafficLights logic

We will now write the complete logic to control all the LEDs in our circuit. This is going to be the first time that we have to move some parts of the code into other functions.

To do that, we start by writing a new function named `trafficLights` that *takes all five LED pins as parameters* and has *no return value*. Inside the function, we start off with an empty, endless loop. Our function should now look like the following:

```
func trafficLights(redLED, greenLED, yellowLED, pedestrianRED,
    pedestrianGreen machine.Pin) {
    for {

    }
}
```

All the logic will be placed inside the `for` loop. The actual logic in the loop consists of two parts:

- Handling signals from the button to stop the traffic and control the pedestrian lights
- Controlling the normal traffic lights flow

We start off with handling the signals from the button. To do that, we check `for stopTraffic` in the `if`, and also have an empty `else` branch. It looks like the following:

```
        if stopTraffic {
    } else {
    }
```

So, when `stopTraffic` is `true`, we want to set our traffic lights phase to be *red*. Also, we want to set the pedestrian lights phase to *green* for 3 seconds and then back to *red* and set `stopTraffic` to `false` afterward, as we handled the signal one time.

Let's implement this logic using the following steps:

1. Set traffic lights phase to red:

    ```
    redLED.High()
    yellowLED.Low()
    greenLED.Low()
    ```

2. Set the pedestrian lights phase to green for 3 seconds:

    ```
    pedestrianGreen.High()
    pedestrianRED.Low()
    time.Sleep(3 * time.Second)
    ```

3. Set the pedestrian lights phase to *red*:

    ```
    pedestrianGreen.Low()
    pedestrianRED.High()
    ```

4. Set `stopTraffic` to `false`, as we have handled the signal:

    ```
    stopTraffic = false
    ```

5. In the `else` block, we just reset the pedestrian lights state to red:

    ```
    pedestrianGreen.Low()
    pedestrianRED.High()
    ```

Okay, that is the part that reacts to `stopTraffic` signals. Underneath that `if-else` block, we are going to implement the actual logic to control the traffic lights flow, which is the same as done earlier. So, we start with the *red* phase, transit to the *red-yellow* phase, then to *green*, then to *yellow*, and then reset *yellow* to be able to start clean again, as follows:

```
redLED.High()
time.Sleep(time.Second)
yellowLED.High()
time.Sleep(time.Second)
redLED.Low()
yellowLED.Low()
greenLED.High()
time.Sleep(time.Second)
greenLED.Low()
yellowLED.High()
time.Sleep(time.Second)
yellowLED.Low()
```

That is all that we have to do in the `trafficLights` function.

Implementing the main logic

Now we only need to run the `trafficLights` function and handle the button input at the same time. This is where **goroutines** come in. As microcontrollers only have one processor core, which works with a single thread, we cannot have real parallel execution of tasks. As we use goroutines on an Arduino UNO, we will need some additional build parameters. We are going to learn about these parameters later, when we flash the program. In our case, we want to have a listener on the button, while still being able to step through the traffic lights process. The logic consists of three steps:

1. Initialize the pedestrian lights with the `red` phase.

2. Run the `trafficLights` function in a goroutine.

3. Handle the button input.

For the first part, we only have to set the `pedestrianRED` LED to `High` and the `pedestrianGreen` LED to `Low`:

```
pedestrianRed.High()
pedestrianGreen.Low()
```

Now we just call `trafficLights` and pass all necessary parameters using a goroutine:

```
go trafficLights(redLED, greenLED, yellowLED, pedestrianRed,
pedestrianGreen)
```

For the last step, we need an endless loop that checks for `buttonInput` and to set `stopTraffic` to `true` if the button is pressed. We also need it to sleep for 50 milliseconds afterward:

```
for {
  if buttonInput.Get() {
    stopTraffic = true
  }
  time.Sleep(50 * time.Millisecond)
}
```

> **Note**
>
> It is necessary to add a sleep time to the loop that handles the button input because the scheduler needs time to run the goroutine. The goroutine is being handled in the time that the main function is sleeping. Also, other blocking functions, such as reading from a channel, can be used to give the scheduler time to work on other tasks.

As we now have completed our logic, it is time to flash the program onto the controller. As we are using goroutines in this project, we need to pass additional parameters to the `tinygo flash` command:

```
tinygo flash -scheduler tasks -target=arduino Chapter02/
traffic-lights-pedestrian/main.go
```

As the ATmega328p has very limited resources, the scheduler is deactivated by default on boards that use this microcontroller. The Arduino UNO is such a board. When using other microcontrollers, we would normally not need to override the default scheduler by setting this parameter.

We have now successfully flashed our program to the Arduino Uno. The traffic lights should start looping all phases and the pedestrian lights should remain in the *red* phase. When clicking the button, the traffic lights should end their loop and then the pedestrian lights should switch to the *green* phase, while the traffic lights remain on the *red* phase for 3 seconds.

> **Note**
>
> Due to the very limited memory on the Arduino Uno, working with goroutines might only work in projects that are not very complex, such as this one.

Summary

We have learned how to build a fully functional traffic lights system with pedestrian lights controlled by a button. We achieved this by building each part of the project separately and assembling it all together at the end.

We learned how to use breadboards, how the color codes on resistors work, why we use resistors when controlling LEDs, and how external LEDs are assembled. Also, we learned how to use push buttons, how to prevent floating signals using pullup resistors, and how to utilize goroutines in TinyGo.

In the next chapter, we are going to learn how to read input from a 4x4 keypad and how to control a servo motor. We are going to utilize this knowledge to build a safety lock that opens when the correct passcode is entered.

Questions

1. Why do we place a resistor between an LED anode and the GPIO port?

2. How do we stop a signal from floating?

3. Why do we sleep after checking a button's state?

4. How would you modify the code to achieve the following behavior?

   a. When the button is pressed, turn off the red and green LEDs of the traffic lights and let the yellow LED blink.

   b. When the button is pressed again: go back to the normal phase rotation.

Further reading

- Resistor Color Conversion Calculator: `https://www.digikey.com/en/resources/conversion-calculators/conversion-calculator-resistor-color-code`

- Goroutines in TinyGo: `https://aykevl.nl/2019/02/tinygo-goroutines`

3
Building a Safety Lock Using a Keypad

We gained basic knowledge of using LEDs, GPIO ports, and resistors in the last chapter. We also learned how to handle input and output. In this chapter, we are going to build a safety lock using a keypad. We will be able to input a passcode in the keypad that triggers a servomotor to unlock a lock. This will be achieved by splitting up the project into individual steps and putting it all together at the end of the chapter.

After working through this chapter, we will know how to write information to the serial port and how to monitor this information. This is a great way to easily debug an application. Then, we are going to write our own driver for a 4x4 keypad, which can be used as passcode input in our case. This 4x4 keypad can also be used as controller input, or as input to start different parts of a program. With that covered, we are going to write the logic to control a servomotor. Servomotors can be used as a lock mechanism and are also often used in remote-controlled planes. In the end, we will have a project where we can set up a passcode, enter the passcode, and trigger the servomotor if the input was correct.

In this chapter, we're going to cover the following main topics:

- Writing to the serial port
- Monitoring the serial port
- Monitoring input from a keypad
- Writing the driver
- Finding drivers for TinyGo
- Controlling a servomotor
- Building a safety lock using a keypad

Technical requirements

We are going to need the following components for this project:

- One Arduino Uno
- One 4x4 membrane keypad
- One SG90 servomotor
- One red LED
- One green LED
- 14 jumper wires
- Two 220 Ohm resistors
- A breadboard

You can find the code for this chapter on GitHub: `https://github.com/ PacktPublishing/Creative-DIY-Microcontroller-Projects-with- TinyGo-and-WebAssembly/tree/master/Chapter03`

The Code in Action video for the chapter can be found here: `https://bit. ly/3uN9OAf`

Writing to the serial port

An easy way to debug your programs on a microcontroller is to write messages to the serial port. You can later use this technique to debug your program, by printing the current step or sensor values, for example.

Let's write a small program to see how writing to a serial port is done. We start by creating a new folder named Chapter03 in the project directory, and inside this new directory, we create another directory named writing-to-serial. Now we have to create a new main.go file and insert an empty main() function. The folder structure should now look like the following:

Figure 3.1 – The folder structure for writing to serial port

Now, follow these steps:

1. We print the word starting followed by a space and print the word program followed by an \n:

```
print("starting ")
print("program\n")
```

2. We endlessly loop, print Hello World, and sleep for a second:

```
for {
    println("Hello World")
    time.Sleep(1 * time.Second)
}
```

3. Now, flash the program to your microcontroller by using the following command:

```
tinygo flash --target=arduino Chapter03/writing-to-
serial/main.go
```

> **Note**
>
> print just writes the text to the serial port and does not insert a character for a newline.
>
> println adds a character for a newline.

Okay, we now have a program on our controller that prints text to the serial port. We have learned about a very convenient way to insert debug logging into our programs. In the next section, we are going to learn how to read data from the serial port on the computer.

Monitoring the serial port

As we are writing debug logs or other messages to the serial port, we need a convenient way to monitor these logs. An easy way to monitor the serial port on all operating systems is to use PuTTy.

Let's first look at how to install PuTTy on various platforms:

- **Linux**: On Linux, PuTTy is available through `apt`. We can install it using the following command:

  ```
  sudo apt install putty
  ```

  Alternatively, we can find `tar.gz` here: `https://www.chiark.greenend.org.uk/~sgtatham/putty/latest.html`

- **MacOS**: On Mac, PuTTy is available using `brew`. We can install it using the following command:

  ```
  brew install Putty
  ```

- **Windows**: On Windows, we can download PuTTy from here: `https://www.putty.org/`. We simply download and run the `.msi` file.

As we have now installed PuTty, it is time to monitor our serial port:

1. Make sure the program from the previous section is flashed on your microcontroller and the USB cable is plugged in.

2. The next step is starting PuTTy. As soon as PuTTy is started, click on **Session** and select **Serial** for **Connection type**. This should look as in the following screenshot:

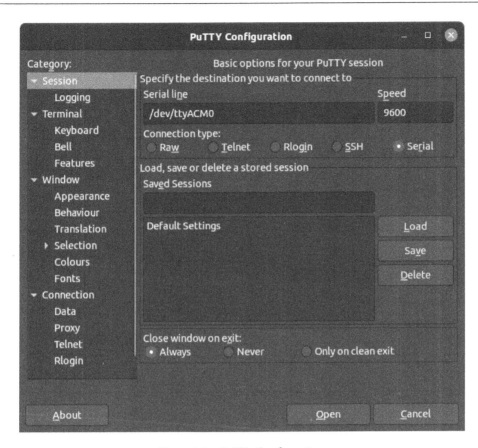

Figure 3.2 – PuTTy Configuration

3. Now we have to choose the serial line. On Windows, this will typically be **COM0**.
    It could also be any other COM port, such as COM5 or similar. Once you have
    found the correct COM port, it will usually stay the same. The Device Manager
    usually lists all devices including the used COM ports in the **Ports (COM & LPT)**
    section. On Linux and Mac, this will typically be /dev/ttyACM0 or /dev/
    ttyUSB0.

4. As we have now successfully configured the session, we can save this configuration. To do so, add `Microcontroller` as the name and click on **Save**. This should look as in the following screenshot:

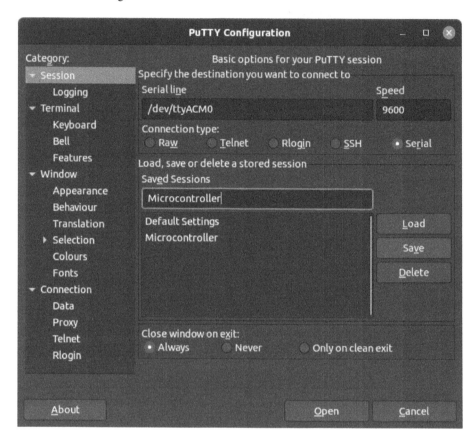

Figure 3.3 – PuTTy saving the configuration

5. As we have now saved the configuration, we can reuse it each time we want to monitor the serial port. Now select **Microcontroller** from the list and click on the **Open** button:

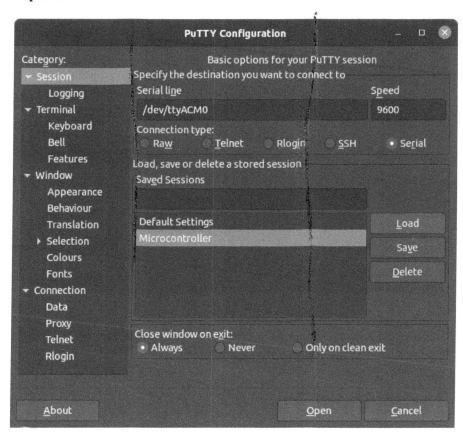

Figure 3.4 – PuTTy Microcontroller session selected

6. After clicking on the **Open** button, a new window opens, which shows the output from our program. It should look similar to the following screenshot:

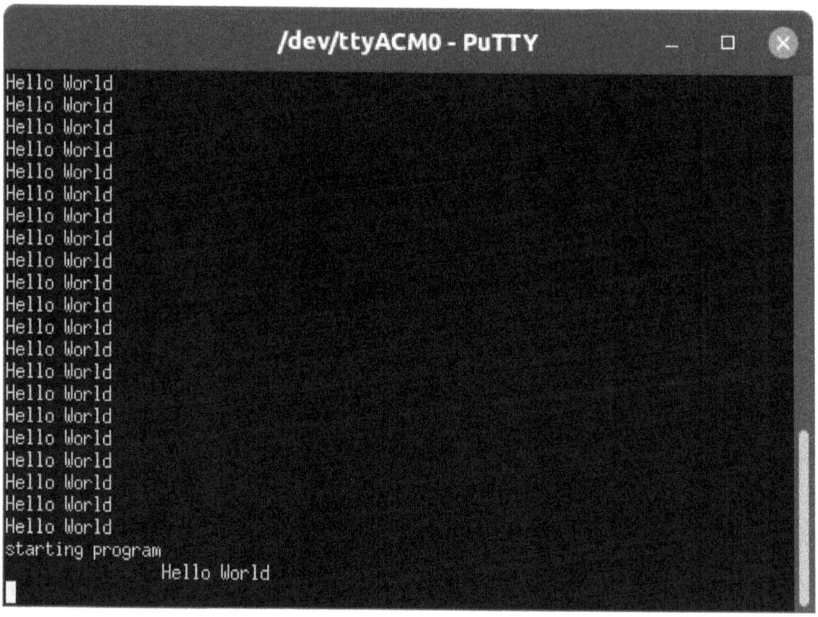

Figure 3.5 – PuTTy program output

We have now learned how to monitor the output of our programs. Next, we are going to learn how to use a 4x4 keypad and monitor button presses.

Monitoring input from a keypad

In this section, we are going to read input from a 4x4 keypad and print the pressed button to the serial port. Since TinyGo does not have a driver for this keypad, we will look at how to create a driver. This will help you understand the process and you can use this knowledge when you need to use other unsupported hardware.

As part of this exercise, I have also followed the process of adding this to the TinyGo codebase and it should be supported in the future. We are going to start by learning how to connect the keypad. Then we will move on to writing a driver, and then we are going to have a brief look at how new drivers are added to TinyGo.

Building the circuit

We start off by assembling the circuit. We are going to need a 4x4 keypad and eight jumper wires. Although we could use jumper cables to directly wire the keypad to the Arduino ports, we are going to wire it through a breadboard. We are going to add more components to this in the upcoming sections. Follow these steps to correctly wire the keypad:

1. Connect pin *D3* to *A32*.

2. Connect pin *D4* to *A31*.

3. Connect pin *D5* to *A30*.

4. Connect pin *D6* to *A29*.

5. Connect pin *D7* to *A28*.

6. Connect pin *D8* to *A27*.

7. Connect pin *D9* to *A26*.

8. Connect pin *D10* to *A25*.

9. Connect *E32* with pin *0* on the keypad.

10. Connect *E31* with pin *1* on the keypad.

11. Connect *E30* with pin *2* on the keypad.

12. Connect *E29* with pin *3* on the keypad.

13. Connect *E28* with pin *4* on the keypad.

14. Connect *E27* with pin *5* on the keypad.

15. Connect *E26* with pin *6* on the keypad.

16. Connect *E25* with P pin in *7* on the keypad.

Having done this, your circuit should look similar to the following screenshot:

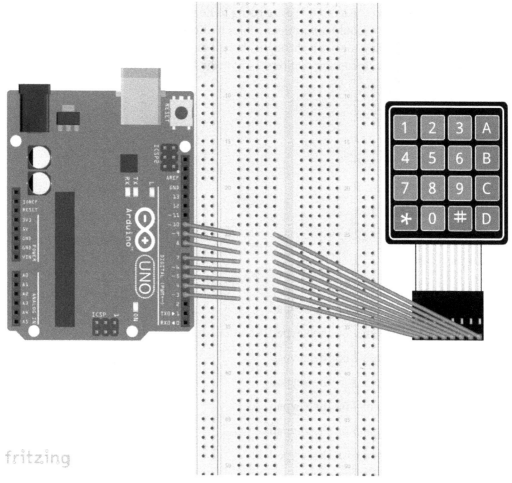

Figure 3.6 – Keypad circuit – image is taken from Fritzing

We have now correctly wired the keypad. Before we can go on with implementing the code, we need to understand how a 4x4 keypad works.

Understanding the workings of a 4x4 keypad

Having a look at the keypad, we realize that it basically consists of **four rows** with **four columns** per row. The keypad comes with eight pins. The first four pins are used for the rows, and the remaining four are used for the columns. To determine which key is being pressed, we just need to find the position of the pressed key in this 4x4 coordinate system.

Button **1**, for example, has the coordinates 0,0 (row 0, column 0), while button **D** has the coordinates 3,3 (row 3, column 3).

In the internal circuit of the keypad, the rows are connected to the columns. When a button is being pressed, the circuit is closed. When the circuit is closed, current flows, which is the signal we can read on a pin. As the keypad is not directly connected to GND and VCC, we need to provide the keypad with power. That is why four pins will be used as input and four will be used as output pins.

I have dissembled such a 4x4 keypad, to provide a visual of the internal circuit:

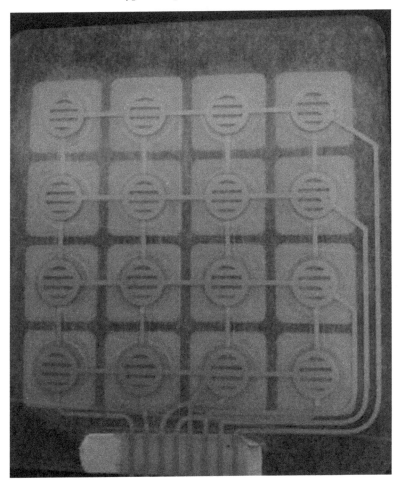

Figure 3.7 – Keypad internal circuit

As we now know that we basically just have to check each coordinate in this 4x4 coordinate system for the correct state, we can go on and write the code.

Writing the driver

As we want to have reusable code, we are going to write a driver package for the keypad. The driver will provide an easy-to-use interface while hiding the more complicated implementation logic. Doing it this way, we can simply reuse the package in later projects even beyond the book. The official TinyGo drivers typically provide a constructor-like function that creates a new instance of the driver and a `Configure` function that takes care of initialization. We are also going to provide a similar API.

Just like in our previous projects, we are going to start by creating a new folder named `controlling-keypad` inside the `Chapter03` folder. Then, we are going to create a `main.go` file with an empty `main` function. Also, we need to create a new folder named `keypad` and create a new file named `driver.go`, and then name the package `keypad`. Your project structure should now look like the following:

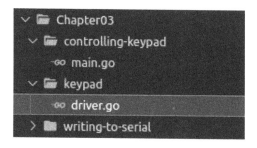

Figure 3.8 – Project structure for writing the driver

We are going to split the logic into the following five parts:

- `Driver` variables
- `Configure`
- `GetIndices`
- `GetKey`
- `main`

Let's understand each of them.

Driver variables

We are going to need some variables inside our `Driver` struct. Follow these steps to set it up:

1. Define a new struct named `Driver`:

    ```
    type Driver struct {
    ```

2. We need an `inputEnabled` variable to debounce the keypresses:

    ```
    inputEnabled bool
    ```

3. `lastColumn` and `lastRow` are used to save the position of the last keypress:

    ```
    lastColumn int
    lastRow int
    ```

4. We need an array of `machine.Pin` to store the column pins:

    ```
    columns [4]machine.Pin
    ```

5. We need an array of `machine.Pin` to store the row pins:

    ```
    rows [4]machine.Pin
    ```

6. We use mapping to map the key values to indices (positions):

    ```
    mapping [4][4]string
    }
    ```

Now we are going to initialize the pins and the `Driver` variables.

Configure

Start off by creating an empty function named `Configure` that takes eight `machine.Pin` function as an argument and is a pointer receiver to `Driver`. This should look like the following snippet:

```
func (keypad *Driver)Configure(r4, r3, r2, r1, c4, c3, c2 ,c1
machine.Pin) {}
```

The next step is to put the initialization logic into this function. To do so, follow these steps:

1. Initialize the column pins using a PinInputPullup config. The internal pullup resistor is going to hold the column to 5 V until a button is being pressed, which we then can read as input:

    ```
    inputConfig := machine.PinConfig{Mode: machine.
                PinInputPullup}
    outputConfig := machine.PinConfig{Mode: machine.
                PinOutput}
    c4.Configure(inputConfig)
    ```

```
c3.Configure(inputConfig)
c2.Configure(inputConfig)
c1.Configure(inputConfig)
```

2. Add the column pins to the `columns` array. By doing so, we can later just use a loop to iterate over all columns:

```
keypad.columns = [4]machine.Pin{c4, c3, c2, c1}
```

3. Initialize the row pins using the `PinOutput` config:

```
outputConfig := machine.PinConfig{Mode: machine.
              PinOutput}
r4.Configure(outputConfig)
r3.Configure(outputConfig)
r2.Configure(outputConfig)
r1.Configure(outputConfig)
```

4. Add all the row pins to the rows array. This enables us to iterate over all the rows using a loop:

```
keypad.rows = [4]machine.Pin{r4, r3, r2, r1}
```

5. Initialize the mapping with the key values. We will be mapping the pressed column and row index to get the correct key value:

```
keypad.mapping = [4][4]string{
    {"1", "2", "3", "A"},
    {"4", "5", "6", "B"},
    {"7", "8", "9", "C"},
    {"*", "0", "#", "D"},
}
```

6. Initialize `inputEnabled`, `lastColumn`, and `lastRow`:

```
keypad.inputEnabled = true
keypad.lastColumn = -1
keypad.lastRow = -1
```

This is everything we need to initialize our program to talk to the keypad.

GetIndices

Now we just need to loop over the arrays and columns and find the pressed key. We start by creating a new function named `GetIndices` that returns two integers and is a pointer receiver to `Driver`. This should look like the following snippet:

```
func (keypad *Driver) GetIndices() (int, int){}
```

Now, follow these steps to implement the function logic:

1. Iterate over all rows:

    ```
    for rowIndex := range keypad.rows {
    ```

2. Set the current `rowPin` to `Low`. We need to do this as we are using internal **pullup resistors**. If **pulldown resistors** were used, we would set `rowPin` to `High` instead:

    ```
    rowPin := keypad.rows[rowIndex]
    rowPin.Low()
    ```

3. Iterate over all columns:

    ```
    for columnIndex := range keypad.columns {
    ```

4. Get the current `columnPin`:

    ```
    columnPin := keypad.columns[columnIndex]
    ```

5. Check whether the current `columnPin` is pressed and execute the logic if we accept input. Disable accepting input and save the current column and row, and then return the indices:

    ```
    if !columnPin.Get() && keypad.inputEnabled {
      keypad.inputEnabled = false
      keypad.lastColumn = columnIndex
      keypad.lastRow = rowIndex
      return keypad.lastRow, keypad.lastColumn
    }
    ```

6. Accept the input again, if the previous key is not pressed anymore:

```
if columnPin.Get() &&
    columnIndex == keypad.lastColumn &&
    rowIndex == keypad.lastRow &&
    !keypad.inputEnabled {
    keypad.inputEnabled = true
}}
```

7. Set `rowPin` to `High` again and close the outer loop:

```
rowPin.High()
}
```

8. Return `-1, -1` if no key was pressed and close the function:

```
return -1, -1
}
```

Calling this function will now tell us the position of the pressed key in the coordinate system. If you want to understand pullup and pulldown resistors in more detail, have a look at the following link: `https://www.electronics-tutorials.ws/logic/pull-up-resistor.html`.

GetKey

Next, we are going to create a function that checks the indices of the pressed key and maps the indices to the key value. To do so, we start with an empty function named `GetKey` that returns a string and is a pointer receiver to `Driver`. This should look like the following snippet:

```
func (keypad *Driver) GetKey() string {}
```

Inside this function, we just call the `GetIndices` method, check whether a button was pressed, and if a button was pressed, we return the key value as a string. This looks like the following:

```
row, column := keypad.GetIndices()
    if row == -1 && column == -1 {
        return ""
    }
return keypad.mapping[row][column]
```

Now, only the `main` logic is missing. Let's look at that next!

main

We call our initialization logic and loop endlessly to check the pressed key. The following steps show how:

1. Initialize `keypadDevice`:

    ```
    keypadDevice := keypad.Driver{}
    keypadDevice.Configure(machine.D3, machine.D4, machine.
        D5, machine.D6, machine.D7, machine.D8, machine.D9,
        machine.D10)
    ```

2. Now, loop endlessly, check for a keypress, and print the value if a key was pressed:

    ```
    for {
      key := keypadDevice.GetKey()
      if key != "" {
        println("Button: ", key)
      }
    }
    ```

Great! That's it. Now we can flash the program and monitor the outputs. Flash the program using the following command:

```
tinygo flash –target=arduino Chapter03/controlling-keypad/main.
go
```

Now, open PuTTy and monitor the serial output while pressing keys on the keypad. The output should look similar to the following screenshot:

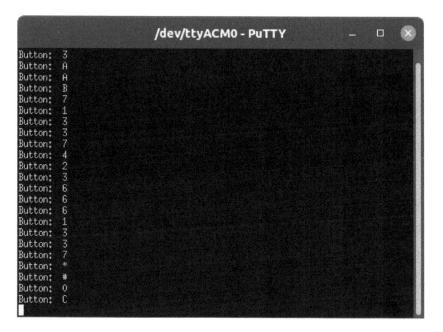

Figure 3.9 – Keypad output in PuTTy

Great, we have successfully written our own driver to monitor button presses on a keypad!

In the next section, we are going to learn where to find TinyGo drivers for peripheral hardware. We are also going to look at the process of contributing to the TinyGo drivers repository.

Finding drivers for TinyGo

As of the time of writing, there are 53 devices supported by TinyGo. The driver we just wrote, which I am going to contribute to TinyGo, will support 54 devices. But where can we find drivers for devices that we want to use? The answer is simple: there is a repository for this purpose. You can find it at `https://github.com/tinygo-org/drivers`.

In the next chapter, we will learn how to use such drivers when using different types of displays.

Contributing drivers to TinyGo

The TinyGo community happily appreciates all contributions. If you develop a driver for a device and want to contribute it to TinyGo, you can follow these simple steps:

1. Open an issue and explain what you want to add and how you plan to implement it.

2. Fork the repository.

3. Create a new branch based on the dev branch.

4. Create a pull request.

You can find the contribution guidelines at the following link: `https://github.com/tinygo-org/drivers/blob/release/CONTRIBUTING.md`.

All in all, my personal experience with the TinyGo community has been extremely positive. They are very polite and will help you out with any kind of problem. I have not encountered a single problem where the community could not give me a helpful answer. Don't be afraid to ask a question in an issue or on the TinyGo channel in the Gophers slack.

> **Note**
> Please do not ask any questions that are directly related to this book in any of the official TinyGo channels, such as Slack or GitHub. If you have any questions regarding this book, you can raise an issue on the accompanying GitHub repository or send an email to me.

As we now know how keypads work and where to find drivers, we can go on with the next part of our safety lock.

Controlling a servomotor

As we are now able to read the input to the keypad, the thing that is missing to build a safety lock is some kind of motor. For that case, we are going to use an SG90 servomotor. As of the time of writing, the timings on the Arduino Uno are not accurate enough to completely control the SG90 servomotor, but that is not a problem for our use case. We are just going to move the servo in one direction, which is clockwise. Also, there is currently no official driver for the SG90 servomotor, so we are going to write our own!

Understanding SG90 servomotors

SG90 servomotors are controlled by **Pulse Width Moduluation** (**PWM**). Basically, the SG90 reads inputs in a 50 Hz period. During this period, we can tell the servomotor to adjust itself to a certain angle by setting a signal for a certain amount of time. The signal length is called the *duty cycle*. After the duty cycle, we wait for the rest of the period. Depending on the duty cycle (the **pulse width**), the SG90 will adjust its angle.

The SG90 can be adjusted to the following three positions:

- 0 degrees (center) using a 1.5 ms pulse
- + 90 degrees (right) using a 2 ms pulse
- - 90 degrees (left) using a 1 ms pulse

It is also possible to adjust the servomotor to all angles in between this by doing some math on the pulse width sizes, but we do not need to do that for our example.

The SG90 typically has three wires:

- Black/brown for ground
- Red for VCC
- Orange/yellow for the PWM signal

Building the circuit

We will build on top of our last example. We just have to add the servomotor by following these steps:

1. Connect the 5 V port from the Arduino Uno to the positive lane on the power bus.
2. Connect a GND port from the Arduino Uno to the ground lane on the power bus.
3. Connect the GND wire from the SG90 to the ground lane on the power bus.
4. Connect the VCC wire from the SG90 to the positive lane on the power bus.
5. Connect the PWM wire from the SG90 to pin *D11* on the Arduino Uno.

Our circuit should now look as in the following screenshot:

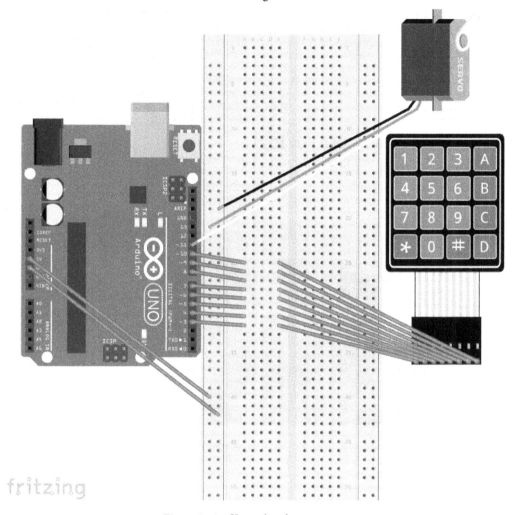

Figure 3.10 – Keypad and servomotor

Excellent. Before we can start programming, we should learn something about PWM pins on the Arduino Uno. Only six pins of the GPIO ports are capable of PWM. The pins are marked with a ~ symbol.

> **Note**
>
> On the Arduino Uno, you can use pins *D3*, *D5*, *D6*, *D9*, *D10*, and *D11* for PWM.

Writing the servo control logic

We need to create a new folder named `controlling-servo` inside the `Chapter03` folder. Next, we create a new `main.go` file inside the new folder and insert an empty `main` function. Also, we need to create a new folder named `servo` with a new `driver.go` file inside the `servo` package. Our project structure should now look like the following:

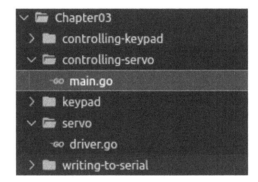

Figure 3.11 – Project structure for servo control logic

> **Note**
>
> PWM is currently being reworked. In the future handling, PWM devices will be much simpler. It is also being handled by hardware PWM instead of emulating the PWM behavior. You can check the progress at the following pull request: `https://github.com/tinygo-org/tinygo/pull/1121`.

The driver that we are now building has the main purpose of teaching us how PWM actually works and is going to work better on all microcontrollers that are not based on the 8-bit AVR architecture, such as the ATmega328P, which is onboard the Arduino Uno. This is due to the fact that the AVR support is still experimental, although it is being improved with nearly every release of TinyGo. As soon as the PR mentioned previously is merged, I recommend using a driver that is based on that hardware PWM support for controlling servos.

Also be aware that as of the time of writing, you manually need to reset the servo when it reaches the rightmost position.

Inside the `driver.go` file, we need to use the following steps to let our servo rotate a bit:

1. Declare package-level constants for the duty cycles and `rightRemainingPeriod`:

    ```
    const centerDutyCycle = 1500 * time.Microsecond
    const centerRemainingPeriod = 18500 * time.Microsecond
    const leftDutyCycle = 2000 * time.Microsecond
    const leftRemainingPeriod = 18000 * time.Microsecond
    const rightDutyCycle = 1000 * time.Microsecond
    const rightRemainingPeriod = 19000 * time.Microsecond
    ```

2. Create a new struct named `Driver` that has `machine.Pin` as a member:

    ```
    type Driver struct {
        pin machine.Pin
    }
    ```

3. Define a new, empty function named `Configure` that takes `machine.Pin` as a parameter and is a pointer receiver to `Driver`:

    ```
    func (servo *Driver) Configure(pin machine.Pin) {}
    ```

4. Configure pin as output:

    ```
    servo.pin = pin
    servo.pin.Configure(machine.PinConfig{Mode: machine.PinOutput})
    ```

5. Loop four times to rotate the motor only about 30 degrees:

    ```
    for position := 0; position <= 4; position++ {
    ```

6. Set a signal for the duty cycle, pull it down, and sleep for the rest of the period:

    ```
    servo.pwm.Pin.High()
    time.Sleep(rightDutyCycle)
    servo.pwm.Pin.Low()
    time.Sleep(rightRemainingPeriod)
    }
    ```

Before we can try out our library, we need to write a small example program. To do so, put the following snippet inside the `main.go` file inside the controlling `servo` folder:

```go
func main() {
    servo := servo.Driver{}
    servo.Configure(machine.D11)
    servo.Right()
}
```

Now we just need to try that program by flashing it using the following command:

```
tinygo flash -target=arduino Chapter03/controlling-servo/main.
go
```

Congratulations, this was the first time that we moved something using code. As we have now learned how to rotate the servo a bit and how to read inputs from a keypad, the next step is to put everything together.

As soon as the refactoring of the PWM is merged to upstream and released in a TinyGo version, you do not want to use the previous driver anymore. For that case, we create a new driver that makes use of hardware PWM instead of emulating the behavior. So go on and create a new folder named `servo-pwm` and create a new `driver.go` file inside. Then follow these steps to implement the better driver:

1. We define the period, which is 20.000 microsecond and create a new `Device` struct, shown as follows:

    ```go
    const period = 20e6
    type Device struct {
        pwm machine.PWM
        pin machine.Pin
        channel uint8
    }
    ```

2. The next step is to add a constructor function as follows:

    ```go
    func NewDevice(timer machine.PWM, pin machine.Pin)
    *Device {
        return &Device{
            pwm: timer,
            pin: pin,
    ```

3. Now we configure the PWM interface. We need to set the period and get the channel for our output pin:

```
func (d *Device) Configure() error {
    err := d.pwm.Configure(machine.PWMConfig{
        Period: period,
    })
    if err != nil {
        return err
    }
    d.channel, err = d.pwm.Channel(machine.Pin(d.pin))
        if err != nil {
            return err}
    return nil
}
```

4. Now we add functions that lets us set the position of the servomotor. We pass in the microseconds for the duty cycles as parameter, as follows:

```
func (d *Device) Right() {
    d.setDutyCycle(1000)
}
func (d *Device) Center() {
    d.setDutyCycle(1500)
}
func (d *Device) Left() {
    d.setDutyCycle(2000)
}
```

5. As a last step, we control the duty cycle of the channel:

```
func (d *Device) setDutyCycle(cycle uint64) {
    value := uint64(d.pwm.Top()) * cycle / (period /
    1000)
    d.pwm.Set(d.channel, uint32(value))
}
```

Let us try how the Set function works. For this, we take a look at the documentation as it the function is very well explained there:

```
// Set updates the channel value. This is used to control the channel duty
// cycle, in other words the fraction of time the channel output is high (or low
// when inverted). For example, to set it to a 25% duty cycle, use:
//
//     pwm.Set(channel, pwm.Top() / 4)
```

Figure 3.12 – The pwm.Set() documentation

Now let us also create an alternative example program that uses the new driver. To do so, create a new folder named controlling-servo-pwm inside the Chapter03 folder and place the following code into the main function:

```
servo := servopwm.NewDevice(machine.Timer1, machine.D9)
err := servo.Configure()
if err != nil {
    for {
        println("could not configure servo:", err.Error())
        time.Sleep(time.Second)
    }
}
for {
    servo.Left()
    time.Sleep(time.Second)

    servo.Center()
    time.Sleep(time.Second)

    servo.Right()
    time.Sleep(time.Second)
}
```

We are using the machine.Timer1 in the preceding example, as the Timer1 is a 16-bit timer, which is usable in combination with the machine.D9 pin. Timer0 and Timer2 are 8-bit timers used by the other PWM pins.

Excellent! I have also added alternative implementations that use hardware PWM based driver instead of the software emulated driver we have in the preceding code, for all following projects in this chapter. You can find them in the GitHub repository in the `Chapter03` folder. I strongly advice to use this implementation of the servomotor driver instead of the one we created first, as this implementation works way better on the Arduino UNO compared to the software emulated PWM driver that we wrote first. Implementing a software emulation of the PWM interface is still a good way to understand how PWM works internally. I have also implemented an alternative program for the final project of this chapter that uses the hardware PWM servo driver. If you cannot build the projects that use the new driver, then the PWM refactoring has not yet made its way onto the TinyGo release branch. But I am very sure, that this feature is going to be released this year (2021).

Building a safety lock using a keypad

We now know how to read input from a keypad and how to control a servomotor. We are going to use this knowledge to build a safety lock that opens when the correct passcode has been entered through the keypad. As we wrote libraries to control the servo and read data from the keypad, we only need to write the logic to check a passcode and light up LEDs. We are going to let the red LED blink each time a key is being pressed. When we enter a wrong passcode, we light up the red LED for 3 seconds. When we enter the correct passcode, we light up the green LED for 3 seconds and trigger the servomotor.

Building the circuit

We are going to reuse the circuits we built in the previous sections of this chapter. As we already have a servo and the keypad wired, we just have to add the LEDs and the resistors.

To build the final circuit, follow these steps:

1. Connect a GND port from the Arduino Uno with the GND lane on the power bus.
2. Place a red LED with the cathode in *G7* and the anode in *G8*.
3. Place a green LED with the cathode in *G11* and the anode in *G12*.
4. Connect *F7* with ground on the power bus using a jumper wire.
5. Connect *F11* with ground on the power bus using a jumper wire.

6. Use a 220 Ohm resistor to connect *D8* with *F8*.

7. Use a 220 Ohm resistor to connect *D12* with *F12*.

8. Connect pin *D12* from the Arduino Uno with *A12* using a jumper wire.

9. Connect pin *D13* from the Arduino Uno with *A8* using a jumper wire.

Our circuit should now look like the circuit in the following screenshot:

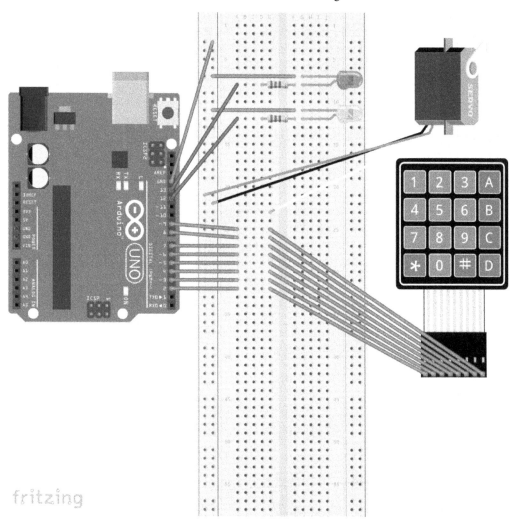

Figure 3.13 – Keypad, servomotor, and LED circuit

Writing the logic

As we have successfully wired our circuit, we can now start to write the logic for our program. We start by creating a new folder named `safety-lock-keypad` inside the `Chapter03` folder and create a new `main.go` file with an empty `main` function inside the new folder. Our project structure should now look like the following:

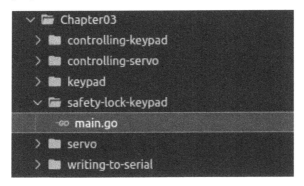

Figure 3.14 – Project structure for safety lock program

As we can reuse our libraries, we just have to concern ourselves with the actual passcode logic. To implement the logic, use the following steps:

1. Import `keypad` and the `servo` driver. Then, you need to adjust the paths to match the paths to the packages in your `Gopath`:

   ```
   "https://github.com/PacktPublishing/Creative-DIY-
   Microcontroller-Projects-with-TinyGo-and-WebAssembly/
   tree/master/Chapter03/keypad"
   ```
   ```
   "https://github.com/PacktPublishing/Creative-DIY-
   Microcontroller-Projects-with-TinyGo-and-WebAssembly/
   tree/master/Chapter03/servo"
   ```

2. Inside the `main` function, we start by initializing `keypadDriver`:

   ```
   keypadDriver := keypad.Driver{}
   keypadDriver.Configure(machine.D2, machine.D3, machine.
       D4, machine.D5, machine.D6, machine.D7, machine.D8,
       machine.D9)
   ```

3. Now, we initialize `servoDriver`:

   ```
   servoDriver := servo.Driver{}
   servoDriver.Configure(machine.D11)
   ```

4. Initialize a new `outPutConfig`:

    ```
    outPutConfig := machine.PinConfig{Mode: machine.
                    PinOutput}
    ```

5. Initialize both LEDs:

    ```
    led1 := machine.D12
    led1.Configure(outPutConfig)
    led2 := machine.D13
    led2.Configure(outPutConfig)
    ```

6. Initialize the passcode with the value `133742`:

    ```
    const passcode = "133742"
    ```

7. Initialize a new variable named `enteredPasscode` with an empty string as the value:

    ```
    enteredPasscode := ""
    ```

8. Read the keypad input:

    ```
    for {
        key := keypadDriver.GetKey()
    ```

9. Check whether a key was pressed and print the pressed key to the serial port, while appending the pressed key to `enteredPasscode`:

    ```
    if key != "" {
        println("Button: ", key)
        enteredPasscode += key
    ```

10. Light up the red LED to provide visual feedback and close the `if` statement:

    ```
            led2.High()
            time.Sleep (time.Second / 5)
            led2.Low()
    }
    ```

11. Check whether `enteredPasscode` has same length as `passcode`:

    ```
        if len(enteredPasscode) == len(passcode) {
    ```

12. If `enteredPasscode` matches the `passcode` value, print `Success` to the serial port , reset the `enteredPasscode` and trigger the servomotor:

```
if enteredPasscode == passcode {
    println("Success")
    enteredPasscode = ""
    servoDriver.Right()
```

13. Light up the green LED to give visual feedback for the success and handle the incorrect passcode case with `else`:

```
    led1.High()
    time.Sleep(time.Second * 3)
    led1.Low()
} else {
```

14. Print `Fail` and the entered passcode to the serial port , that helps us, when debugging the program and also reset the `enteredPasscode`:

```
    println("Fail")
    println("Entered Password: ", enteredPasscode)
    enteredPasscode = ""
```

15. Light up the red LED to give visual feedback for the failure and close the `else` and `if` cases:

```
    led2.High()
    time.Sleep(time.Duration(time.Second * 3))
    led2.Low()
}
}
```

16. Sleep for 50 milliseconds and close the `for` loop. This helps to debounce the keypresses:

```
time.Sleep(50 * time.Millisecond)
}
```

Great, we have now written the complete logic for our final project in this chapter. Now flash the program using the following command:

```
tinygo flash –target=arduino Chapter03/safety-lock-keypad/main.
go
```

As we have now successfully flashed the program, open PuTTy and open the **Microcontroller** serial session by loading your saved profile. Now enter a random passcode to let the program fail. The red LED should light up for 3 seconds and the output in PuTTy should look as in the following screenshot:

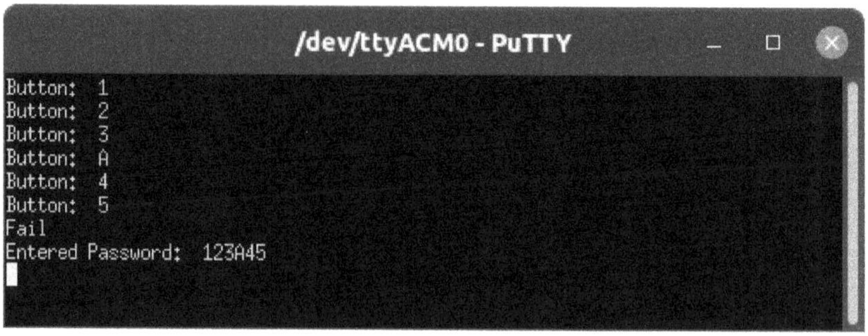

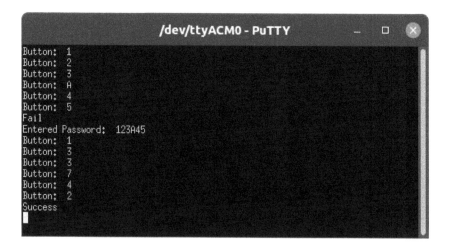

Figure 3.15 – Incorrect input

Now, let's try the correct passcode, so enter `133742` as the passcode. The output should now look similar to the following screenshot:

Figure 3.16 – Correct input

Excellent, we have successfully built a circuit that accepts a passcode and triggers a servomotor when the correct passcode has been entered.

You can find a alternative implementation that uses the new refactored PWM jere: `https://github.com/PacktPublishing/Creative-DIY-Microcontroller-Projects-with-TinyGo-and-WebAssembly/blob/master/Chapter03/safety-lock-keypad-pwm/main.go`

Summary

In this chapter, we have learned how to write messages to the serial port and how to configure PuTTy to monitor messages on the serial port. We have then used this knowledge to output keypresses on a keypad that we controlled using a driver that we wrote. During that procedure, we learned how to write drivers for devices that currently have no official drivers and also learned about the contribution process of the driver's repository from TinyGo.

Then we learned how to control a servomotor and wrote a library to do so. As the last step, we combined everything we learned in this chapter to build a safety lock that accepts a passcode to open up the lock. This knowledge can be very useful if you ever want to build a door lock or a flight control system, where you need to control servomotors. The keypad can also be used as a gamepad, where you use the keys as input. As a bonus, we also wrote two drivers that we can reuse in all upcoming projects after finishing the book.

In the next chapter, we are going to learn how to read sensor values using ADC pins, how to find thresholds in values, how to control a pump, and how to use relays and buzzers.

Questions

1. Having learned about the coordinate system we used for the keypad, what are the coordinates for key 3?

2. In our final project, we checked whether the input is correct when the correct passcode length has been reached. How would you change the code to get it to check whether the passcode is correct when the key number has been pressed?

4
Building a Plant Watering System

In the previous chapters, we learned how to write to the serial port and how to monitor the serial port on our computers. Furthermore, we learned how to write drivers for components, which have not yet been implemented by the TinyGo community, and we used this knowledge to write a driver for a 4x4 keypad and a servo motor in *Chapter 3, Building a Safety Lock Using a Keypad*.

We are now going to build on top of this knowledge in this chapter by introducing a new type of pin and we are going to build an automated plant watering system using some new devices. We will be able to pump water from a container into a plant's soil, measure the soil's moisture, check the water level of a container, and let a buzzer beep when the water level in the container is below a certain threshold. This will be achieved by splitting the project up into single steps and putting it all together at the end of the chapter.

After working through this chapter, we will know how to read input from analog pins, how to measure thresholds in sensor data, how to let a buzzer beep, and how to control a pump using relays.

In this chapter, we're going to cover the following main topics:

- Reading soil moisture sensor data
- Reading water level sensor data
- Controlling a buzzer
- Controlling a pump
- Watering your plants

Technical requirements

We are going to need the following components for this project:

- An Arduino UNO
- Capacitive Soil Moisture Sensor v1.2
- K-0135 Water Level Sensor
- Passive buzzer with 2 pins
- Micro submersible water pump DC 3V-5V
- Breadboard power supply module
- Jumper wires
- One breadboard
- One 100 Ohm resistor

These components can usually be found in online stores and also in local electronic supply stores. Most components used in this book are also part of so-called **Arduino Starter Kits**.

You can find the code for this chapter on GitHub: `https://github.com/ PacktPublishing/Creative-DIY-Microcontroller-Projects-with- TinyGo-and-WebAssembly/tree/master/Chapter04`

The Code in Action video for the chapter can be found here: `https://bit. ly/3tlhRnx`

Reading soil moisture sensor data

When automatically watering plants, we need to know when we have to add water to the soil. An easy way to detect that the soil is too dry is to use a soil moisture sensor. We are going to use a capacitive soil moisture sensor in this project, which provides the readings as an analog signal.

The sensor has the following technical specifications:

- A 3.3 V to 5.0 V supply range

- A 3.3 V operating range

- An analog output in the range of 1.5 V to 3.3 V

- An operating current of 5 mA

Sensors from other manufacturers might differ slightly in these specs. Datasheets are usually provided by the vendor you buy the hardware from. We'll now start off by assembling the circuit.

Assembling the circuit

We only need some cables, the sensor itself, and a breadboard to begin with. Depending on the manufacturer of the sensor, the labels on the port of your sensor might differ. The one I use has the following labels:

- AOUT (short for **Analog out**)

- VCC (+) (short for **Voltage Common Collector**)

- GND (-) (stands for **Ground**)

Now assemble the circuit as per the following list:

1. Connect the *GND* port to *GND* on the power bus using a jumper wire.

2. Connect port *D2* to *A1* on the breadboard using a jumper wire.

3. Connect port *A5* to *A2* on the breadboard using a jumper wire.

4. Connect *E1* on the breadboard to *AOUT* on the sensor.

5. Connect *E2* to *VCC* on the sensor.

6. Connect *GND* from the power bus with *GND* on the sensor.

Your circuit should now look similar to the following figure:

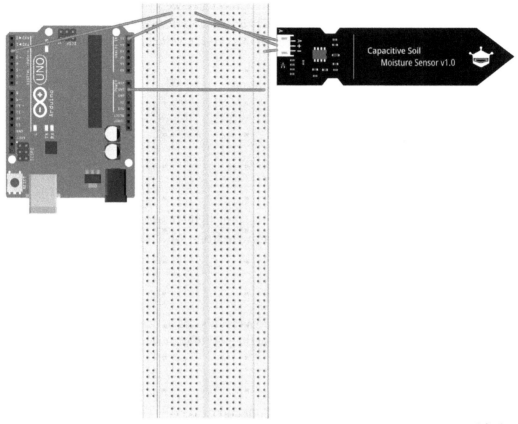

Figure 4.1 – Soil Sensor Circuit image taken from Fritzing

Great! We have successfully assembled the circuit. We are going to use this circuit to create a small sample project to read values from the sensor.

Finding thresholds

Our next task is to find out the values that indicate the following states:

- Dry
- In water

To check the dryness, we need to create a new folder for this project.

Start off by creating a new folder named `Chapter04`. Inside this folder, create a new folder named `soil-moisture-sensor-thresholds` and inside this folder, create a new `main.go` file and insert an empty `main()` function. The folder structure should now look like the following:

Figure 4.2 – Folder structure for Soil moisture sensor threshold

Now follow these steps:

1. Import the `machine` package as follows:

    ```
    import "machine"
    ```

2. Initialize the registers needed for ADC:

    ```
    machine.InitADC()
    ```

3. Create a new variable named `soilSensor` of the type `machine.ADC` with `Pin` `machine.ADC5`:

    ```
    soilSensor := machine.ADC{Pin: machine.ADC5}
    ```

4. Configure the pin so it is able to read analog values:

    ```
    soilSensor.Configure()
    ```

5. Configure the machine D2 pin as output and set it to high. We do not store this inside a new variable, as we will never change the state of the pin again. We only use it to provide currency:

    ```
    machine.D2.Configure(machine.PinConfig{Mode: machine.
    PinOutput})
    machine.D2.High()
    ```

6. Read the sensor value two times a second in an endless loop and print it to the serial port:

```
for {
    value := soilSensor.Get()
    println(value)
    time.Sleep(500 * time.Millisecond)
}
```

Great, we have successfully written our first program that reads sensor data from an analog pin. Now we need to ascertain threshold values. To do so, first flash the program onto your Arduino using the following command:

```
tinygo flash –target=arduino Chapter04/soil-moisture-sensor-
thresholds/main.go
```

Now open up PuTTY and select the microcontroller profile to see the sensor readings. This should look like the following screenshot:

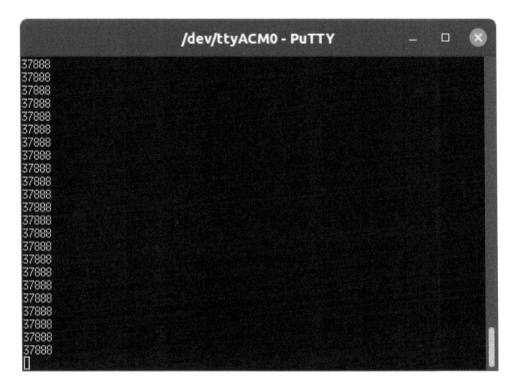

Figure 4.3 – Soil moisture sensor output in PuTTY

The value is pretty stable at **37888**. You might notice some small changes in the value between the readings. Just take the highest value you see in that case.

We are now going to declare **37888** as the threshold for dry values. So, everything equal or above this value can be considered completely dry. The values that you receive from your sensor might differ a bit, so you can just do the same; have a look at the values, and take the lowest one as your threshold.

> **Note**
>
> Take care that your sensor is completely dry and clean. Otherwise, you might get in trouble with your dry value.

Excellent! We just managed to find out the value for completely dry soil. Now we need to find a value for being completely wet (in water).

Now take a glass of water and put the sensor in there, while watching the sensor readings in PuTTY. **Be really careful that you only put the sensor so deep in the water that it reaches the white line on it!** Do not let the water touch the electronics above! Check the following figure for this:

Figure 4.4 – Capacitive Soil Moisture Sensor in a glass of water

Let's have a look at the sensor readings in PuTTY – you can find them in the following figure:

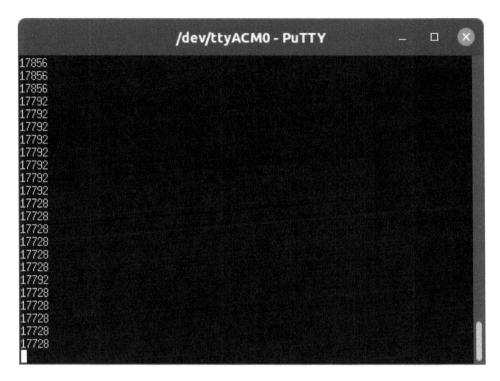

Figure 4.5 – Soil sensor readings from inside a glass of water in PuTTY

This time we take the highest value we can find in PuTTY as our threshold, which in my case is **17856**. We have used GPIO pins in all previous chapters, but we have not yet used the **Analog Digital Converter**, so let's understand how the **Analog Digital Converter (ADC)** works on Arduino before we continue to write a library for the sensor.

Understanding ADC in TinyGo

The Arduino UNO ADC has 10-bit precision. The value returned by the Get() function is of the type uint16. So, internally, the Get() function tells the ADC to scale the 10-bit value to a 16-bit value.

In general, we can use the following equation to get an ADC result:

```
Resolution of the ADC / System Voltage = ADC Value / Analog
Voltage measured
```

As we know that the Arduino UNO has 10-bit precision and the voltage is about 5V, we can insert this into the equation to get the following:

```
1023 / 5V = ADC Value / Analog Voltage Measured
```

Let's say the analog voltage measured is 3.33V. This will result in the following:

```
1023 / 5V = ADC Value / 3.33V
```

Now we do some math equation magic and get this:

```
1023 / 5V * 3.33V = ADC Value
That resolves to: ADC Value = 681
```

This result will now scale to 16 bits, which equals a bit shift left by 6 bits.

The result we would get from TinyGo would be like the following:

```
ADC Value = 43584
```

Since we have now discovered how ADC works, we can now continue to write a small library that will help us use the sensor later on.

Writing a library for the sensor

As having reusable code is a nice thing, we will now go on to write a small library, to reuse it in the last part of this chapter, *Watering your plants.*

To do so, we need to create a new folder named soil-moisture-sensor inside the Chapter04 folder. In our newly created folder, we create a new empty driver.go file and name the package soil. The structure should now look like the following:

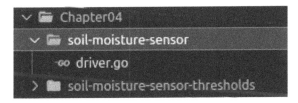

Figure 4.6 – Folder structure for soil moisture sensor library

We want to have an interface that provides a function to get the current MoistureLevel, instance which will be an enum-like type. Also, we want to provide functionality to turn the sensor on and off, so it does not draw current all the time.

To achieve that, perform the following steps:

1. Define a new interface named `SoilSensor` with `Get()`, `Configure()`, `On()`, and `Off()` functions:

    ```
    type SoilSensor interface {
        Get() MoistureLevel
        Configure()
        On()
        Off()
    }
    ```

2. Define a new struct named `soilSensor`. This struct is going to contain the pin that is being used to turn the sensor on and off and the pin that is being used to read the sensor value. Also, we want to be able to configure thresholds that are used to identify whether the sensor is completely dry or is in water:

    ```
    type soilSensor struct {
    ```

3. Add members that save the thresholds for `completelyDry` and `water`:

    ```
    waterThreshold          uint16
    completelyDryThreshold  uint16
    category                uint16
    pin                     machine.Pin
    adc                     machine.ADC

    voltage                 machine.Pin
    }
    ```

4. Define an enum-like type so we can easily check for these values when using the library. We chose six `MoistureLevel` categories here, to have a clear distinction between the different states of the soil:

    ```
    type MoistureLevel uint8
    const (
        CompletelyDry MoistureLevel = iota
        VeryDry
        Dry
        Wet
        VeryWet
        Water

    )
    ```

5. Define a constructor function that takes `waterThreshold`, `dryThreshold`, `dataPin`, and `voltagePin` and returns `SoilSensor`:

```
func NewSoilSensor(waterThreshold, dryThreshold, dataPin,
voltagePin machine.Pin) SoilSensor {
```

6. Use the thresholds and create `category`, which will later be used to calculate the `category` value. As we want to have six categories, we divide the values by six, which have been read from where the sensor lies:

```
category := (dryThreshold - waterThreshold) / 6
```

7. Set all values and return a pointer to a new instance of `soilSensor`:

```
return &soilSensor{
    waterThreshold: waterThreshold,
    completelyDryThreshold: dryThreshold,
    category: category,
    pin: dataPin,
    voltage: voltagePin,
}
}
```

8. Define a new `func` named `Get`, which is in a function receiver to a pointer to `soilSensor` and returns `MoistureLevel`:

```
func (sensor *soilSensor) Get() MoistureLevel {
```

9. Read the value from `sensor` and save it the new variable value of type `float32`:

```
value := sensor.adc.Get()
```

10. Check whether the value is greater than or equal to `completelyDryThreshold`:

```
switch {
case value >= sensor.completelyDryThreshold:
    return CompletelyDry
```

11. Check whether `value` falls into the second category:

```
case value >= sensor.completelyDryThreshold-sensor.
category:
    return VeryDry
```

12. Check whether `value` falls into the third category:

```
case value >= sensor.completelyDryThreshold-sensor.
category*2:
    return Dry
```

13. Check whether `value` falls into the fourth category:

```
case value <= sensor.completelyDryThreshold-sensor.
category*3:
    return Wet
```

14. Check whether `value` falls into the fifth category:

```
case value >= sensor.completelyDryThreshold-sensor.
category*Ł4:
    return VeryWet
```

15. The only remaining possible state is `Water`, so we use the default case here:

```
default:
    return Water
    }
}
```

16. Define a function named `Configure` that has a function receiver for a pointer of `soilSensor`. We use a pointer receiver as we set values on the `soilSensor` instance, which we would otherwise lose outside of this function scope:

```
func (sensor *soilSensor) Configure() {
```

17. Configure `dataPin` for ADC usage:

```
sensor.adc = machine.ADC{Pin: sensor.pin}
sensor.adc.Configure(machine.ADCConfig{}
```

18. Configure the `voltage` pin as output and set it to `Low`:

```
sensor.voltage.Configure(machine.PinConfig{Mode:
machine.PinOutput})
sensor.voltage.Low()
}
```

19. Add a function to turn on the voltage:

```
func (sensor *soilSensor) On() {
    sensor.voltage.High()
}
```

20. Add a function to turn off the voltage:

```
func (sensor *soilSensor) Off() {
    sensor.voltage.Low()
}
```

This is the complete logic we need for our library. Let's test our code in the next section.

Testing the library

Next, we will write an example to test the new library. To do so, we need to create a new folder named `soil-moisture-sensor-example` inside the `Chapter04` folder and create a `main.go` file with an empty `main()` function inside. Your project structure should now look like the following:

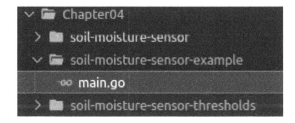

Figure 4.7 – Testing the soil moisture sensor library

To test our new library, follow these steps:

1. Import the `machine`, `time`, and `soil-moisture-sensor` packages as shown in the following code. Note that the path to your library will be a bit different, depending on where it lies on your filesystem:

```
import (
    "machine"
    "time"
    "github.com/PacktPublishing/Creative-DIY-
Microcontroller-Projects-with-TinyGo-and-WebAssembly/
    Chapter04/soil-moisture-sensor"
)
```

2. Inside the `main` function, initialize the ADC interface:

```
machine.InitADC()
```

3. Create a new instance of `SoilSensor`. The values in this example differ slightly from the ones measured in the last example. These values tend to trigger the `Water` and `CompletelyDry` states earlier:

```
soilSensor := soil.NewSoilSensor(18000, 34800, machine.
            ADC5, machine.D2)
```

4. Now we call the `Configure` function, which initializes our pins:

```
soilSensor.Configure()
```

5. Start an endless loop and turn on the sensor, and wait a brief moment to let the readings stabilize a bit:

```
for {
    soilSensor.On()
    time.Sleep(75 * time.Millisecond)
```

6. Then switch over the result from the `Get()` function and print a string depending on the case:

```
        switch soilSensor.Get() {
        case soil.CompletelyDry:
            println("completely dry")
        case soil.VeryDry:
```

```
            println("very dry")
        case soil.Dry:
            println("dry")
        case soil.Wet:
            println("wet")
        case soil.VeryWet:
            println("very wet")
        case soil.Water:
            println("pure water")
    }
```

7. Turn the sensor off again and wait a second until the next reading starts:

```
        soilSensor.Off()
        time.Sleep(time.Second)
    }
}
```

As we now have the complete code to test our library, let's flash it onto our Arduino and let's check the output in PuTTY. Use the following command to flash it:

```
tinygo flash -target=arduino Chapter04/soil-moisture-sensor-
example/main.go
```

The best thing to do now is to actually put the sensor in very dry soil while having an eye on the readings. Then add some water, to check whether you will see the Wet, Very Wet, and maybe the Water states. Before we continue with the next section, we should definitely check it. If the readings seem odd, try to adjust the thresholds, which are handled in the library in the NewSoilSensor function.

> **Note**
>
> Keep in mind that you could harm plants by pouring too much water into the soil. Also, it is not necessary to stick the sensor all the way down into the soil until it reaches the white line. I suggest leaving some air between the white line and the soil so you have some buffer between the electronics and the soil. When I did my testing, I achieved good results when having about 1 cm of air as a buffer.

We have now learned how to calibrate the Capacitive Soil Moisture sensor, used the ADC interface for the first time, and written a new library. Using this library, we are able to tell the humidity state of the soil. In the next section, we are going to learn how to use a water level sensor.

Reading water level sensor data

As we plan to have a water tank later in the chapter, it will be beneficial to have a water level sensor, so we can tell when the tank is empty. We'll start off by adding the sensor to our existing circuit. Follow these steps to do so:

1. Connect pin *A4* from the Arduino with *F22* on the breadboard using a jumper cable.
2. Connect pin *D3* from the Arduino with *F21* on the breadboard using a jumper cable.
3. Connect *J22* on the breadboard with the *S* (Signal) port on the sensor using a jumper cable.
4. Connect *J21* on the breadboard with the + (*VCC*) port on the sensor using a jumper cable.
5. Connect - *GND* from the sensor with *GND* on the power bus.

The result should now look like the following figure:

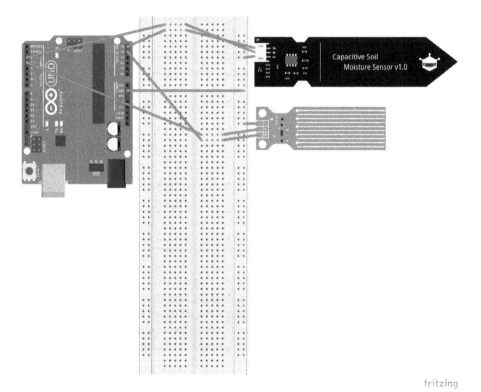

Figure 4.8 – Water level sensor – image is taken from Fritzing

After assembling this, we can continue to also create a small library for this sensor.

Writing a water level sensor library

There are many different types of water level sensors. The cheap ones that are often part of Arduino Starter Kits often suffer from corrosion. To prevent that, we are going to also add the possibility to turn it on and off. But first, we are going to have a look at the technical data:

- **Operating voltage**: 5 V

- **Operating current**: Less than 20 mA

- **Working temperature**: 10° to 30°

So, having this sensor draws less than 20 mA current. We can again use a *GPIO* pin to power it.

We start by creating a new folder named `water-level-sensor` inside the `Chapter04` folder. Inside the new folder, create a new file named `driver.go` and name the package `waterlevel`. The folder structure should now look like the following:

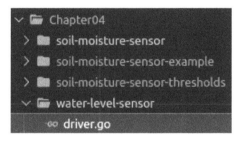

Figure 4.9 – Folder structure for water level sensor library

As the project structure is now set up, we can go on to implement the actual library. Just follow these steps:

1. Import the `machine` package:

    ```
    import "machine"
    ```

2. Define a new interface named `WaterLevel` with the following functions. The functions are going to be explained when we implement them:

    ```
    type WaterLevel interface {
        IsEmpty() bool
        Configure()
        On()
        Off()
    }
    ```

3. Define a struct named `waterLevel` with the following members:

```
type waterLevel struct {
    dryThreshold uint16
    pin machine.Pin
    adc machine.ADC
    voltage machine.Pin
}
```

4. Define a new constructor-like function that takes `dryThreshold`, `dataPin`, and `voltagePin`:

```
func NewWaterLevel(dryThreshold uint16, dataPin,
voltagePin machine.Pin) WaterLevel {
    return &waterLevel{
        dryThreshold: dryThreshold,
        pin: dataPin,
        voltage: voltagePin,
    }
}
```

5. Add the `IsEmpty` check. We'll just check whether the sensor reading is lower than our threshold:

```
func (sensor *waterLevel) IsEmpty() bool {
    return sensor.adc.Get() <= sensor.dryThreshold
}
```

6. Configure the sensor pin for ADC usage and configure the voltage pin as output:

```
func (sensor *waterLevel) Configure() {
    sensor.adc = machine.ADC{Pin: sensor.pin}
    sensor.adc.Configure(machine.ADCConfig{})
    sensor.voltage.Configure(machine.PinConfig{
        Mode: machine.PinOutput,
    })
    sensor.voltage.Low()
}
```

7. Turn the power on:

```
func (sensor *waterLevel) On() {
    sensor.voltage.High()
}
```

8. Turn the power off:

```
func (sensor *waterLevel) Off() {
    sensor.voltage.Low()
}
```

We have now written the library for the water level sensor.

Testing the library

Now let's write a small example program to test our library. To do so, we start by creating a new folder named `water-level-sensor-example` inside the `Chapter04` folder. Inside the new folder, create a new `main.go` file with an empty `main` function inside. The folder structure should now look like the following:

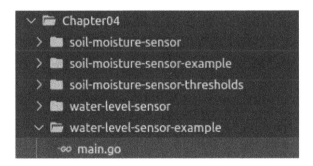

Figure 4.10 – Folder structure for testing the library

As the project structure is now set up, we can go on to write the test code. To do so, follow these steps:

1. Initialize the ADC interface:

```
machine.InitADC()
```

2. Create a new instance of `WaterLevelSensor`:

```
waterLevelSensor := waterlevel.NewWaterLevel(7000,
                        machine.ADC4, machine.D3)
```

3. Configure the pins:

```
waterLevelSensor.Configure()
```

4. We turn the sensor on and then wait a brief moment to let the readings of the sensor stabilize before we access it. Then, print the result of IsEmpty() every second:

```
for {
    waterLevelSensor.On()
    time.Sleep(100 * time.Millisecond)
    println("tank is empty", waterLevelSensor.IsEmpty())
    waterLevelSensor.Off()
    time.Sleep(time.Second)
}
```

When the water level sensor does not touch any water, the returned value should be 0. We chose 7000 as dryThreshold in this case, so the tip of the sensor can be inside the water and still be able to tell us it's empty. That will be useful later on, in the case when we also need to pump water. This is for when the pump should not run when there is not enough water to pump. We should play around with this threshold value a bit. Do so by flashing the program to your Arduino, check the water presence with the sensor, and when it realizes that there is water, change the threshold value and flash again.

Flash the program by using the following command:

```
tinygo flash –target=arduino Chapter04/water-level-sensor-
example/main.go
```

So we now have written a library that checks whether any kind of water tank is empty. In the next section, we are going to use a buzzer to have an audio signal when the water tank is empty.

Controlling a buzzer

We are going to write a very simple buzzer library. We only want the buzzer to make any sound, regardless of the pitch. We start off by adding the buzzer to the circuit. To do so, follow these steps:

1. Connect *D4* from the Arduino to *A31* on the breadboard using a jumper wire.

2. Use a *100* Ohm resistor to connect *E31* with *G31* on the breadboard.

3. Connect the *VCC* pin from the buzzer with *J31*.

4. Connect the *GND* pin to *GND* on the power bus.

The circuit should now look like the following figure:

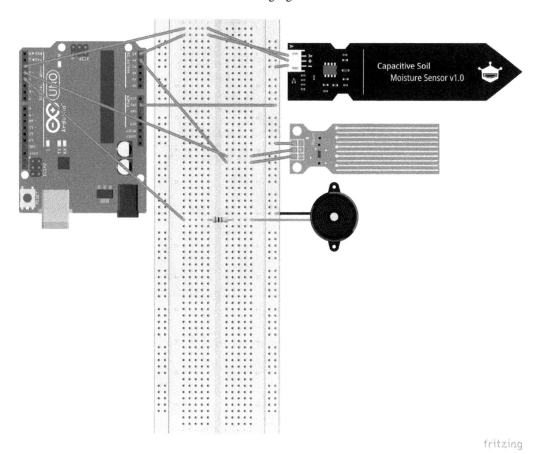

Figure 4.11 – Buzzer – image taken from Fritzing

As we have now added the buzzer to the circuit, we can now start to write our library.

Writing a buzzer library

The buzzer library will have two functions: Configure(), which sets up the pin, and the Beep() function, which will make the sound.

We start off by creating a new folder named buzzer inside the Chapter04 folder. Inside the new folder, create a file named driver.go and name the package buzzer. The project structure should now look like the following:

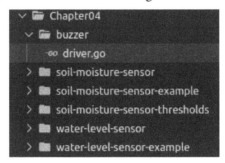

FIgure 4.12 – Project structure for buzzer library

Now follow these steps to implement the driver:

1. Define an interface named Buzzer that has a Configure function and a Beep function:

    ```go
    type Buzzer interface {
        Configure()
        Beep(highDuration time.Duration, amount uint8)
    }
    ```

2. Create a struct named buzzer that holds machine.Pin:

    ```go
    type buzzer struct {
        pin machine.Pin
    }
    ```

3. Add a function named NewBuzzer that returns Buzzer:

    ```go
    func NewBuzzer(pin machine.Pin) Buzzer {
        return buzzer{pin: pin}
    }
    ```

4. Add a function named Configure that configures pin as output:

    ```go
    func (buzzer buzzer) Configure() {
        buzzer.pin.Configure(machine.PinConfig{
            Mode: machine.PinOutput,
        })
    }
    ```

5. Define a function named `Beep`, which takes `time.Duration` and `amount` of the `uint8` type as parameters:

```
func (buzzer buzzer) Beep(highDuration time.Duration,
amount uint8) {
```

6. Loop the amount of times and let the buzzer beep and sleep in between:

```
for i := amount; i > 0; i-- {
    buzzer.pin.High()
    time.Sleep(highDuration)
    buzzer.pin.Low()
    time.Sleep(highDuration)
}
}
```

That is all for the buzzer library.

Now we are going to test the library with a small example project.

To do so, we first create a new folder named `buzzer-example` inside the `Chapter04` folder. Inside the new folder, create a new `main.go` file with an empty `main()` function in it. The project structure should now look like the following:

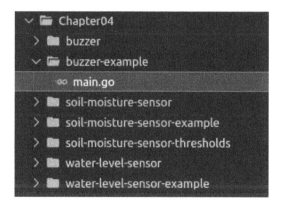

Figure 4.13 – Testing the buzzer

Now put the following inside the `main` function:

1. Get a new instance of `buzzer` and configure it:

```
buzzer := buzzer.NewBuzzer(machine.D4)
buzzer.Configure()
```

2. Loop forever, beep three times for `100` milliseconds, and then sleep for 3 seconds:

```
for {
    buzzer.Beep(time.Millisecond*100, 3)
    time.Sleep(3 * time.Second)
}
```

That's all the code we need to test the buzzer. To try this example, flash it by using the following command:

```
tinygo flash -target=arduino Chapter4/buzzer-example/main.go
```

When the program runs, you should be able to hear the buzzer. If it does not start to make a sound after a brief amount of time, check all the cables and pins again.

We have now successfully written a very simple buzzer library and tested it using an example project. In the next section, we are going to control a pump.

Controlling a pump

As pumps tend to draw more current than simple sensors, we are not going to power the pump directly through a *GPIO* port. Drawing too much current could permanently damage the Arduino. So, we will use an external power supply and a relay to power the pump. Before we start assembling the circuit, let's have a brief look at how relays work.

Working with relays

A relay that is used for microcontroller projects typically comes mounted on a board, which typically has six ports. It has three input ports: *VCC*, *GND*, and *Signal*. It also has three output ports: *normally open*, *common*, and *normally closed*.

When a *high signal* is given, the current flows between *normally open* and *common*.

When a *low signal* is given, the current flows between *normally closed* and *common*.

As we now know how to use a relay, we can continue to add the new components to our circuit. To do so, follow these steps:

1. Connect the *GND* pin from the relay to *GND* on the power bus using a jumper wire.

2. Connect the *VCC* pin from the relay to *VCC* on the power bus using a jumper wire.

3. Connect the *Signal* (in) pin from the relay to *D5* on the Arduino using a jumper wire.

4. Connect the *VCC* pin of the pump to the *normally open* port on the relay.

5. Connect the *GND* pin of the pump to *GND* on the power bus.

6. Connect the *common* pin of the relay with the *VCC* lane on the power bus.

7. Connect *VIN* from the Arduino to *VCC* on the power bus using a jumper wire.

Your circuit should now look similar to the one in the following figure:

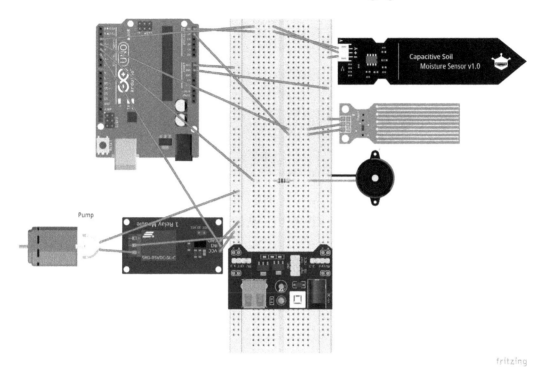

Figure 4.14 – Full circuit including a pump – image taken from Fritzing

With this circuit, we will be able to power the Arduino using an external power supply. As we might want to water plants that are not anywhere near a USB port, we have connected the *VIN* pin on the Arduino with the *VCC* lane on the power bus, which is powered by our external power supply. We will now go on and write a library that is able to control the pump.

Writing a pump library

The pump library is basically going to have two functions: `Configure`, which sets up the pin, and the `Pump` function, which pumps for a given duration and a given number of iterations.

We start by creating a new folder named pump inside the Chapter04 folder. Inside the new folder, create a driver.go file and name the package pump. The project structure should now look like the following:

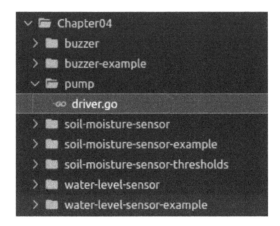

FIgure 4.15 – Project structure for pump library

Now, as we have set up the project structure, we can go on to write the code to control the pump. To do so, follow these steps:

1. Define the interface named Pump, which has the Configure or Pump functions:

    ```
    type Pump interface {
        Configure()
        Pump(duration time.Duration, iterations uint8)
    }
    ```

2. Define a new struct named pump that holds machine.Pin:

    ```
    type pump struct {
        pin machine.Pin
    }
    ```

3. Define a function named NewPump that takes machine.Pin and returns a new pointer to pump:

    ```
    func NewPump(pin machine.Pin) Pump {
        return &pump{
            pin: pin,
        }
    }
    ```

4. Then define a function named `Configure` and set `pin` as the output pin:

```
func (pump *pump) Configure() {
    pump.pin.Configure(machine.PinConfig{
    Mode: machine.PinOutput,
    })
}
```

5. Next, define a function named `Pump` and loop it `iterations` times, to set `pin` to high, sleep for `duration`, and set it back to `low` again:

```
func (pump *pump) Pump(duration time.Duration, iterations
uint8) {
    for i := iterations; i > 0; i-- {
        pump.pin.High()
        time.Sleep(duration)
        pump.pin.Low()
        time.Sleep(duration)
    }
}
```

This was everything we need for our pump library. We could now go on and create a small example project to test the library. To do so, we'll create a new folder named `pump-example` inside the `Chapter04` folder and create a `main.go` file with an empty `main` function inside. The project structure should now look like the following:

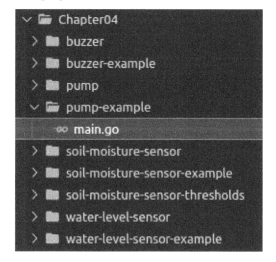

Figure 4.16 – Testing the pump

Inside the `main` function, we add the following:

1. Create a new instance of `pump` by calling the `NewPump` function and hand in `machine.D5` as `pin`:

    ```
    pump := pump.NewPump(machine.D5)
    pump.Configure()
    ```

2. Loop forever and `pump` 3 times for `350` milliseconds and `sleep` for `30` seconds afterward, as follows:

    ```
    for {
        pump.Pump(350*time.Millisecond, 3)
        time.Sleep(30 * time.Second)
    }
    ```

This is the complete example code to try out our pump.

> **Note**
>
> Due to the laws of physics being the way they are, I would recommend that a possible receiving container should always be placed above the upper water level of the source container since the water will continue to flow even though the pump stopped pumping.

Now put your pump in a glass of water or some other water tank and try it out by flashing the program using the following command:

```
tinygo flash –target=arduino Chapter04/pump-example/main.go
```

Use this example to find out good pump duration and iteration times that don't pump too much water. Keep in mind that we want to water plants, so this is going to help us find good values.

This was the last component we needed. We have learned how to use a relay to power and control a pump and we wrote a new library. We are now going to put everything together in the next section.

Watering your plants

We are now going to utilize every component we created in the past sections. Putting everything together, we will be building a completely automated plant watering system.

To start off, we need to create a new folder named `plant-watering-system` inside the `Chapter04` folder. Inside the new folder, create a new `main.go` file with an empty `main()` function inside. The final project structure should now look like the following:

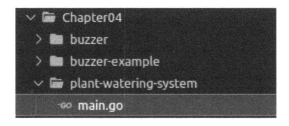

Figure 4.17 – Project structure for plant watering system

Now, inside the `main` function, follow these steps:

1. Initialize the `ADC` interface:

    ```
    machine.InitADC()
    ```

2. Initialize a new `soilSensor`:

    ```
    soilSensor := soil.NewSoilSensor(18000, 34800, machine.
                  ADC5, machine.D2)
    soilSensor.Configure()
    ```

3. Initialize a new `waterLevelSensor`:

    ```
    waterLevelSensor := waterlevel.NewWaterLevel(7000,
                        machine.ADC4, machine.D3)
    waterLevelSensor.Configure()
    ```

4. Initialize a new pump:

    ```
    pump := pump.NewPump(machine.D5)
    pump.Configure()
    ```

5. Initialize a new `buzzer`:

```
buzzer := buzzer.NewBuzzer(machine.D4)
buzzer.Configure()
```

6. Turn `waterLevelSensor` on and sleep for a brief amount so the readings can stabilize:

```
for {
    waterLevelSensor.On()
    time.Sleep(100 * time.Millisecond)
```

7. Check whether the water container is empty, turn the sensor off, beep 3 times, and then sleep for an hour, before continuing the `for` loop:

```
    if waterLevelSensor.IsEmpty() {
        waterLevelSensor.Off()
        buzzer.Beep(150*time.Millisecond, 3)
        time.Sleep(time.Hour)
        continue
    }
```

8. If the water container is not empty, turn off `waterLevelSensor`:

```
waterLevelSensor.Off()
```

9. Turn on `soilSensor` and sleep for a brief amount of time to let the readings stabilize:

```
soilSensor.On()
time.Sleep(100 * time.Millisecond)
```

10. Then switch over the result of `soilSensor.Get()`:

```
switch soilSensor.Get() {
```

11. If the soil is `VeryDry` or `CompletelyDry`, turn off the soil sensor and pump water:

```
case soil.VeryDry, soil.CompletelyDry:
    pump.Pump(350*time.Millisecond, 3)
```

12. In all other cases, turn off `soilSensor` and sleep for an hour:

```
default:
    soilSensor.Off()
    time.Sleep(time.Hour)
    }
    }
```

This is everything we need for this final project. You can try out the program by flashing it onto your Arduino using the following command:

```
tinygo flash -target=arduino Chapter04/plant-watering-system/
main.go
```

> **Important note**
>
> Keep in mind that every plant has other needs in terms of water. So, we will need to tweak the values for the amount of water pumped to fit the needs of the plant being watered.

We have now successfully built a complete automated plant watering system and flashed it onto the Arduino.

Summary

In this chapter, we learned how to read sensor values using the ADC interface. We also learned how the ADC interface translates voltage to digital values, and then we utilized this knowledge to write the `soil moisture sensor` library.

We then wrote the `water level sensor` library by utilizing the knowledge we gathered in the first project of this chapter. Then we learned how to use a **buzzer** and wrote a very simple library that enables us to let a buzzer create warning sounds. After that, we learned how relays work and utilized this knowledge to control a **pump** using a library we wrote. At the end of this chapter, we put all the libraries in a single project and only had to add a small amount of control logic to build the automatic plant watering system.

In the next chapter, we are going to learn how to use supersonic sensors and how to control seven-segment displays.

Questions

1. Why are the water level sensor and soil moisture sensor not permanently powered?

2. When is the circuit between *normally open* and *GND* closed in a relay?

References

The Capacitive Soil Moisture Sensor fritzing part was part of a collection from the following repository: `https://github.com/OgreTransporter/fritzing-parts-extra`

5
Building a Touchless Handwash Timer

In *Chapter 4*, *Building a Plant Watering System*, we learned how the ADC interface works, and we used that knowledge to write libraries for a capacitive soil moisture sensor and a water level sensor. We also wrote a small library to control a buzzer and learned how relays work and used that knowledge to control a pump using our code. Then we used all this knowledge to build an automated plant watering system.

In this chapter, we are going to build a touchless handwash timer. After working through this chapter, you will know how ultrasonic sound sensors work and how to measure distance with them. We are going to utilize this knowledge to create a sensor that recognizes a hand that is between 20 and 30 centimeters away from the sensor to start a timer. The timer will then be displayed on a 7-segment display. While implementing this, we will also learn about the MAX7219 chip and how to use it to control different display types.

In this chapter, we're going to cover the following main topics:

- Introducing the Arduino Nano 33 IoT

- Measuring distance

- Using a 7-segment display

- Putting it all together

Technical requirements

We are going to need the following components for this project:

- An Arduino Nano 33 IoT

- An HC-SR04 sensor

- An external power supply module

- HS420561K 4-Digit 7-segment display common cathode

- A MAX7219 or MAX7221 serial input/output common-cathode display driver

- One 10,000 Ohm resistor

- One 1,000 Ohm resistor

- One 2,000 Ohm resistor

- 2 breadboards

- Jumper wire cables

Most of the components are part of the so-called Arduino Starter Kit. If you do not have such a set, they can be acquired at any electronics supply store.

You can find the code for this chapter on GitHub at `https://github.com/PacktPublishing/Creative-DIY-Microcontroller-Projects-with-TinyGo-and-WebAssembly/tree/master/Chapter05`

The Code in Action video for the chapter can be found here: `https://bit.ly/3e2IYgG`

Introducing the Arduino Nano 33 IoT

We have reached a point where the TinyGo support for the Arduino UNO has reached its limit. At the time of writing, it is not possible to resolve this shortcoming for the current and following chapters using an Arduino UNO. The cause of this is a missing **Serial Peripheral Interface** (**SPI**) implementation for the Arduino UNO. However, we will be able to overcome this in the near future, as there is a `Pull` request for that opened by me. Additionally, the **Alf and Vegard's RISC** (**AVR**) backend in the TinyGo compiler toolchain has some problems in the versions used by the current TinyGo version, and the code won't compile. So, let's take a look at another board, which is fully supported by TinyGo—the Arduino Nano 33 IoT. Compared to UNO, the Nano 33 IoT is a powerhouse. Here are its technical specifications:

- **Microcontroller**: AMD21 Cortex®-M0+ 32bit low power ARM MCU
- **Radio module**: U-blox NINA-W102
- **Operating voltage**: 3.3V
- **Input Voltage (limit)**: 2V
- **DC Current per I/O Pin**: 7 mA
- **Clock speed**: 48 MHz
- **CPU Flash Memory**: 256 KB
- **SRAM**: 32 KB
- **GPIO Pins**: 14
- **Analog input pins**: 8 (8/10/12 bit)
- **Analog output pins**: 1 (10 bit)

So, the Arduino Nano 33 IoT has a much higher clock speed, more RAM, and more flash memory while operating on 3.3V instead of 5V. Additionally, on top of that, the Nano 33 IoT is capable of Wi-Fi communication.

Now, let's take a brief look at the 5V output capabilities of the Arduino Nano 33 IoT.

> Note
>
> Although the Arduino Nano 33 IoT has a 5V output pin, this pin is deactivated by default. To activate that pin, some soldering needs to be done.
>
> When powering the Arduino Nano 33 IoT through USB, we also have a 5V current available on the `Vin` pin, but that pin is intended to power the Arduino. We will be handling devices that require 5V input, but that is not a problem; we are just going to use an external power supply to power these devices.

Now that we have a brief understanding, let's take a look at the pinout. The following diagram shows the pinout of the Arduino Nano 33 IoT:

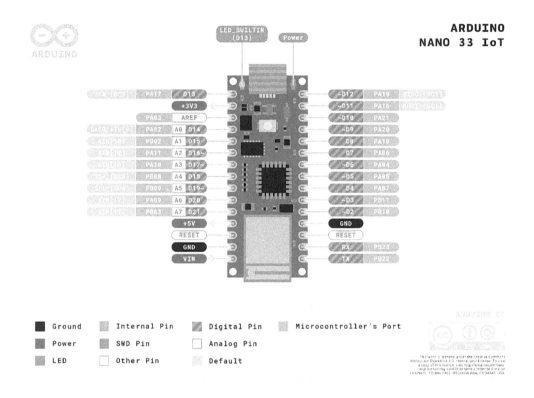

Figure 5.1 – The Arduino Nano 33 IoT pinout

The source of the pinout diagram can be found at `https://store.arduino.cc/arduino-nano-33-iot`.

In this section, we took a brief look at the technical specs of the Arduino Nano 33 IoT. However, before we can use it in our project, we need to install some dependencies.

Installing Bossa

Bossa is needed in order to flash programs onto the Arduino Nano 33 IoT.

First, let's take a look at the installing process on a Mac system:

- You can simply install the dependencies using the following command:

```
brew cask install bossa
```

To install the dependencies on Windows download, execute the following `msi`:

```
https://github.com/shumatech/BOSSA/releases/
download/1.9.1/bossa-x64-1.9.1.msi
```

- When executing the `msi`, choose the following installation path:

```
c:\Program Files
```

Now, add `bossa` to the path using the following command:

```
set PATH=%PATH%;"c:\Program Files\BOSSA";
```

- In order to install `bossa` on a Linux system, execute the following commands:

```
sudo apt install libreadline-dev libwxgtk3.0-gtk3-dev
git clone https://github.com/shumatech/BOSSA.git
cd BOSSAmakesudo
cp bin/bossac /usr/local/bin
```

To verify the installation success, use the following command:

```
bossac -help
```

Up-to-date information on how to install the needed dependencies can be found at `https://tinygo.org/microcontrollers/arduino-nano33-iot/`.

We have now set up the dependencies required to flash programs on the Arduino Nano 33 IoT. Let's move on to the first project of this chapter.

Learning to measure distances

If you have ever wondered how touchless soap dispensers or touchless blow dryers register that there is a hand beneath them, there is a good chance that they are using the HC-SR04 ultrasonic sensor. We are going to use this sensor to measure the distance between an object and the sensor. Let's begin with the HC-SR04 sensor.

Understanding the HC-SR04 sensor

The HC-SR04 sensor emits an ultrasound at 40k Hz, which travels through the air and bounces back if the emitted pulse collides with any object in its path. The sensor cannot be used as a detector for other ultrasound pulses, as it only registers echoes from the exact same pulse that it itself emitted. Typically, these sensors look similar to the one in the following photograph:

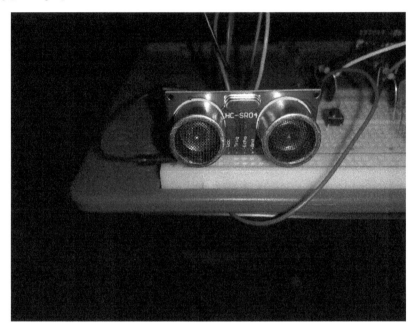

Figure 5.2 – The HC-SR04 sensor

This sensor has the following technical specifications:

- It has a detection range from 2 to 400 centimeters.

- It draws less than 2 mA current.

- It has a working voltage of 5V.

- It has a resolution of 0.3 centimeters.

- It has an angle of fewer than 15 degrees.

The sensor has the following three ports:

- *VCC*: This is used to power the sensor.

- *TRIG*: This triggers the pulse.

- *ECHO*: This receives the echo of the pulse.

Now, let's take a look at how exactly an ultrasonic pulse can be used to measure the distance between the sender and an object. The sensor emits eight pulses that travel through the air. If they hit an object, they get reflected and travel back as an echo, as shown in the following diagram:

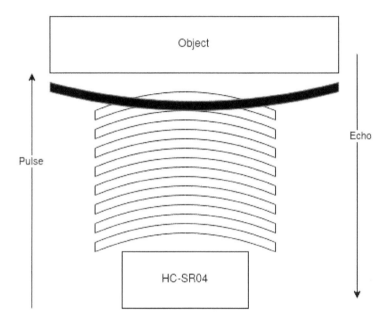

Figure 5.3 – Eight pulses and one echo

When recognizing the echo, the **echo** pin will be set to high from the sensor for the exact same time frame that the pulse needed to leave and return to the sensor. Now, we only need some math to calculate the distance.

The pulse travels at 340 m/s, which is the speed of sound in the air. This can also be expressed as 0.034 m/μs (microseconds). If the object is about 30 centimeters away from the sensor, the pulse needs to travel for about 882 microseconds. The **echo** pin will be set to high for exactly as long as the pulse needs to travel the entire path; that is why we need to divide the result by 2. And as the last step, we are going to divide the travel time by 0.034 to get the traveled distance in centimeters.

Here is how this example works out:

```
Time = Distance / Speed
t = s/v
t = 30cm / 0.034m/us
t = 882.352941176us
```

Let's rearrange that formula to get `distance`:

```
Distance = Time * Speed
30cm = 882.352941176us * 0.34m/us
```

Now we have learned how to use an ultrasonic sound sensor, theoretically, we can now go on to prove this theory by trying it out in real time.

Assembling the circuit

Before we start to assemble the circuit, we need to make sure that the echo pin from the sensor, which will be connected to an input pin on the Arduino, does send 3.3V signals instead of 5V signals. For this, we can make use of a **voltage divider**. The echo pin of the sensor outputs 5V, but the Arduino Nano 33 IoT should not be connected to 5V, as this could cause permanent damage to the Arduino. That is why we make use of the voltage divider.

The formula to calculate the **output voltage (Vout)** is as follows:

```
Vout = Vs * R2 / (R1 + R2)
```

Here, `Vs` is the source voltage, `R2` is the resistor that is connected to the source voltage, and `R1` is the resistor that is connected to the ground.

So, we are going to need a 2,000-Ohm resistor for R2 and a 1,000-Ohm resistor for R1. This is going to result in the following equation:

3.333V = 5 * 2,000 / (1,000 + 2,000)

Now that we have learned how to build a voltage divider, we can go ahead and assemble the circuit using the following steps:

1. Place the HC-SR04 sensor on the breadboard with the *VCC* pin in the *J* row on the breadboard.
2. Connect the *VCC* lane on the power bus with *VCC* of the sensor using a jumper wire.
3. Connect the *GND* lane on the power bus with *GND* of the sensor using a jumper wire. Connect *D2* from the Arduino with *Trig* of the sensor using a jumper wire.
4. Use a 2,000-Ohm resistor to connect `GND` with *C53* on the breadboard using a jumper wire.
5. Use a 1,000-Ohm resistor to connect *Echo* with *A53* on the breadboard jumper wire.

6. Now, connect *D3* from the Arduino with *B53* on the breadboard using a jumper wire. We can read the 3.3V *Echo* signal here.

7. Place an external power supply on the breadboard. Take care to **set the jumpers to 5V**.

This is everything we need to write and test a library. Your circuit should now look similar to the following diagram:

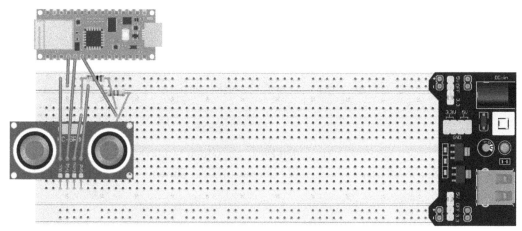

Figure 5.4 – The HC-SR04 circuit (image taken from Fritzing)

We have now learned how an ultrasonic sound sensor works and have assembled a circuit. As has been the case with the previous chapters, here, we also start by creating a library to control the sensor.

Writing a library

We are going to write a library that has a function that returns the current distance from the sensor to an object, or 0 if the object is out of range. We start by creating a new folder, called Chapter05, inside our project. Inside the new Chapter05 folder, create a new folder, called ultrasonic-distance-sensor, and create a new driver.go file. Name the package hcsr04. Your folder structure should look like the following:

Figure 5.5 – The project structure for writing a library

Now that we have set up our project structure, we can start to write the actual logic. To do this, perform these steps:

1. Define a new constant at the package level, name it speedOfSound, and set its value to 0.0343, which is the speed of sound in centimeters per microsecond:

    ```
    const speedOfSound = 0.0343
    ```

2. Next, define a new interface and call it HCSR04, as shown in the following code:

    ```
    type Device interface {
            Configure()
            GetDistance() uint16
            GetDistanceFromPulseLength(
                float32) uint16
    }
    ```

3. Then, we define a new struct, called hcsr04, which holds the trigger and echo pins as well as a timeout in microseconds, as shown in the following code:

    ```
    type device struct {
            trigger machine.Pin
            echo machine.Pin
            timeout int64
    }
    ```

4. Next, we add a function named NewHCSR04, which takes a trigger and echo pin along with the maxDistance in centimeters and returns HCSR04:

    ```
    func NewDevice(trigger, echo machine.Pin, maxDistance
        float32) HCSR04 {
    ```

5. Calculate the timeout in microseconds. We multiply the maxDistance by 2 because the pulse needs to travel to an object and back again. We then divide the result by speedOfSound:

    ```
    timeout := int64(maxDistance * 2 / speedOfSound)
    ```

6. Create a new instance of `hcsr04`, set `trigger`, `echo`, and `timeout`, and return a pointer to the new instance:

```
return &device{
        trigger: trigger,
        echo: echo,
        timeout: timeout,
    }
}
```

7. Add the `Configure` function, which is a pointer receiver that configures `trigger` as output and the `echo` pin as input:

```
func (sensor *device) Configure() {
    sensor.trigger.Configure(
        machine.PinConfig{Mode: machine.PinOutput},
    )
    sensor.echo.Configure(
        machine.PinConfig{Mode: machine.PinInput},
    )
}
```

8. Add the `sendPulse` function, which pulls the `trigger` high for `10` microseconds and then sets the `trigger` to low again. This will trigger eight ultrasonic pulses in the HC-SR04 sensor:

```
func (sensor *device) sendPulse() {
    sensor.trigger.High()
    time.Sleep(10 * time.Microsecond)
    sensor.trigger.Low()
}
```

9. Add a new function named `GetDistance`, which returns `uint16` and is a pointer receiver. First, the function sends out a pulse and then listens for the echo. We receive an incoming echo when the echo pin reads a high value:

```
func (sensor *device) GetDistance() uint16 {
    i := 0
    timeoutTimer := time.Now()
```

```
sensor.sendPulse()
for {
    if sensor.echo.Get() {
        timeoutTimer = time.Now()
        break
    }
    i++
```

10. Check whether i is greater than 15. We do this to save some time, as comparing an integer is a very fast operation in comparison to getting a current timestamp. If the time since our timer started is greater than our configured timeout, then return 0, which we can use as the timeout value:

```
if i > 15 {
    microseconds := time.Since(timeoutTimer).
        Microseconds()
    if microseconds > sensor.timeout {
        return 0
    }
}
}
```

Now we have to measure the time when the echo pin is set to high:

```
var pulseLength float32
i = 0
for {
    if !sensor.echo.Get() {
        microseconds := time.Since(timeoutTimer).
            Microseconds()
        pulseLength = float32(microseconds)
        break
    }

    i++
    if i > 15 {
        microseconds := time.Since(timeoutTimer).
            Microseconds()
```

```
        if microseconds > sensor.timeout {
            return 0
        }

    }
}
return sensor.GetDistanceFromPulseLength(pulseLength)
}
```

11. Add a function named GetDistanceFromPulseLength, which takes pulseLength as a parameter, returns the distance in centimeters, and is a pointer receiver:

```
func (sensor *hcsr04) GetDistanceFromPulseLength(
    pulseLength float32) uint16 {
```

12. As the pulseLength parameter is the time that the signal needs to travel to the target and back, we need to divide it by 2:

```
pulseLength = pulseLength / 2
```

13. To get the result in centimeters, we need to multiply pulseLength with speedOfSound:

```
result := pulseLength * speedOfSound
```

14. Return result as uint16, as we don't care about decimal places:

```
return uint16(result)
}
```

This is all the code we need for the library. From now on, we can use this library to measure distances using an HCSR-04 sensor.

Before we move on to test the library in a real-world example, let's use the GetDistanceFromPulseLength function to take a brief look at how unit testing can be done in TinyGo.

Unit testing in TinyGo

TinyGo does support unit testing, which comes in handy when you have complicated logic and do not want to flash every single change onto your microcontroller when you are trying to find a bug. Let's take a look at what is currently is supported in a practical way. To do this, create a new file, called `driver_test.go`, inside the `ultrasonic-distance-sensor` folder and name the package `hcsr04_test`. The project structure should look similar to the following:

Figure 5.6 – The project structure for the first unit test

Now, let's add our first unit test in TinyGo. To do so, perform these steps:

1. Add a new function named `TestGetDistanceFromPulseLength_30cm`:

   ```
   func TestGetDistanceFromPulseLength_30cm(t *testing.T)
   {
   ```

2. Create a new `HCSR04` instance. We do not really need a parameter for this test, as these parameters will not be used; however, we can add some correct ones anyway:

   ```
   sensor := hcsr04.NewHCSR04(
       machine.D2, machine.D3, 100)
   ```

3. Calculate the `distance` for the given `pulseLength` parameter, which is exactly the length of a pulse for a distance of 30 centimeters:

   ```
   distance := sensor.GetDistanceFromPulseLength(
       1749.27113703)
   ```

4. Check whether the `distance` equals `30`. If it does not, we fail the test and log some information, as follows:

   ```
   if distance != 30 {
       t.Error("Expected distance: 30cm", "actual
           distance: ", distance, "cm")
   }
   }
   ```

This was our first unit test in TinyGo. Do not forget to import the `testing` package. Just like in the normal program code, we can make use of the standard Golang package. You can run the test by using the following command:

```
tinygo test --tags "arduino_nano33" Chapter05/ultrasonic-
distance-sensor/driver_test.go
```

TinyGo internally makes heavy use of **build flags**. These build flags, such as `arduino_nano33`, are used to decide which packages and files are needed to build the current code. The test will not compile if we omit the `-tags` parameter, as the `machine` package would then be missing.

The output of the test should look like the following:

```
=== RUN    TestGetDistanceFromPulseLength_30cm
--- PASS:  TestGetDistanceFromPulseLength_30cm
```

Now we know that we can make use of very simple tests to test our logic in TinyGo. Let's go one step further and do a **table-driven test**. Perform these steps:

1. The next step is to add a new function, called `TestGetDistanceFromPulseLength_TableDriven`, right underneath the other test. The code snippet looks like the following:

   ```
   func TestGetDistanceFromPulseLength_TableDriven(
       t *testing.T) {
   ```

2. Add four test cases, each of them with a `Name`, the expected `Result`, and the `PulseLength`, which we use as input, as follows:

   ```
   var testCases = [4]struct {
       Name string
       Result uint16
       PulseLength float32
   }{
       {
           Name: "1cm",
           Result: 1,
           PulseLength: 58.8235294117},
       {
           Name: "30cm",
   ```

```
        Result: 30,
        PulseLength: 1749.27113703},
    {
        Name: "60cm",
        Result: 60,
        PulseLength: 3498.54227405},
    {
        Name: "400cm",
        Result: 400,
        PulseLength: 23323.6151603},
}
```

3. Create a new instance of the `HCSR04 Device` instance. It should look like the following snippet:

```
sensor := hcsr04.NewDevice(
        machine.D2, machine.D3, 100)
```

4. Now we can run a test for each `testCase` in the array. This looks like the following code:

```
for _, testCase := range testCases {
    t.Run(testCase.Name, func(t *testing.T) {
```

5. Calculate the `distance`. And check whether we get a result that differs from the predefined test cases:

```
distance := sensor.GetDistanceFromPulseLength(
    testCase.PulseLength)
if distance != testCase.Result {
    t.Error("Expected distance:", testCase.Name,
        "actual distance: ", distance, "cm")
}
})
}
}
```

This was everything for the tests. Now, let's run the tests again using the following command:

```
tinygo test --tags "arduino_nano33" Chapter5/ultrasonic-
distance-sensor/driver_test.go
```

The output should now look like the following:

```
tinygo test --tags "arduino nano33" Chapter05/ultrasonic-distance-sensor/driver test.go
=== RUN    TestGetDistanceFromPulseLength 30cm
--- PASS: TestGetDistanceFromPulseLength 30cm
=== RUN    TestGetDistanceFromPulseLength TableDriven
=== RUN    TestGetDistanceFromPulseLength TableDriven/1cm
    --- PASS: TestGetDistanceFromPulseLength TableDriven/1cm
=== RUN    TestGetDistanceFromPulseLength TableDriven/30cm
    --- PASS: TestGetDistanceFromPulseLength TableDriven/30cm
=== RUN    TestGetDistanceFromPulseLength TableDriven/60cm
    --- PASS: TestGetDistanceFromPulseLength TableDriven/60cm
=== RUN    TestGetDistanceFromPulseLength TableDriven/400cm
    --- PASS: TestGetDistanceFromPulseLength TableDriven/400cm
--- PASS: TestGetDistanceFromPulseLength TableDriven
PASS
```

Figure 5.7 – The tinygo test output

> **Note**
>
> Since building binaries for Windows is currently not supported, the preceding
> tinygo test command is going to fail on a Windows systems. One option
> for Windows users is to use the WSL for unit tests. Another possibility is to
> set a build target by using the -target parameter. Windows does support
> building the **wasm** or **wasi** targets, but as our code depends on the machine
> package, that will not work for this specific test. This is because the machine
> package is not available for the wasm and wasi targets.

Now we know that we can also use table-driven tests in TinyGo. At the time of writing, the majority of the testing package seems to be implemented. Currently, only the Helper() function seems to not be implemented. However, there could be one or two small things that I have not found yet, which might not work. Additionally, we have checked that our logic to calculate the distance seems to be working as expected.

With that covered, we can go on and write a small example program to test the rest of our code on real hardware.

Writing an example program for the library

We have now checked that our formula to calculate the distance from a pulse length input seems to be correct. So, we can move ahead and create an example that outputs the measured distance to serial. To do that, first, we need to create a new folder, called `ultrasonic-distance-sensor-example`, inside the `Chapter05` folder. Additionally, we need to create a new `main.go` file with an empty `main` function. The project structure should look similar to the following:

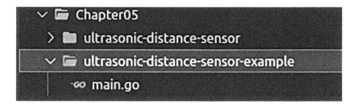

Figure 5.8 – The project structure for the example program

The example logic consists of initializing the sensor and then printing the distance each second. All of this is done inside the `main` function. It looks like the following snippet:

```
sensor := hcsr04.NewHCSR04(machine.D2, machine.D3, 80)
sensor.Configure()

for {
    distance := sensor.GetDistance()
    if distance != 0 {
        println("Current distance:", distance, "cm")
    }
    time.Sleep(time.Second)
}
```

This is the complete code for the example. The library was imported with an alias named `hcsr04`. Now, let's flash the program onto the Arduino Nano 33 IoT using the following command:

```
tinygo flash --target=arduino-nano33 Chapter05/ultrasonic-distance-sensor-example/main.go
```

To check the output, we can use the same PuTTY profile that we created in *Chapter 3, Building a Safety Lock Using a Keypad*. Open up PuTTY and select the **Microcontroller** profile. Make sure that you have the USB cable plugged into the same port as the Arduino UNO beforehand. Depending on the current distance from the sensor to any object, the output should look similar to the following screenshot:

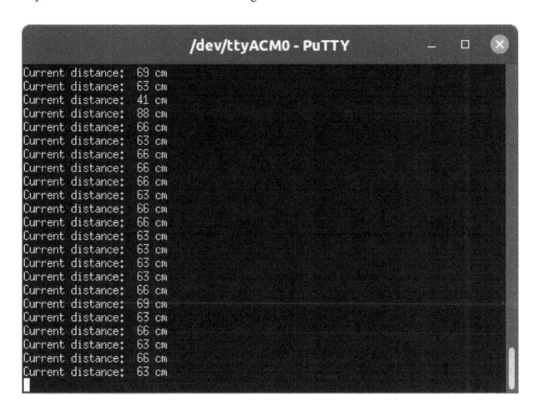

Figure 5.9 – Sensor readings in PuTTy

In this section, we wrote a library for the HC-SR04 sensor, learned that unit testing also works in TinyGo, and then wrote an example project to test the library. So, we are now able to measure distance, which is the first half of our project.

In the next section, we will look at 7-segment displays, as we need 7-segment displays to display a timer in our final project.

Using 4-digit 7-segment displays

A 7-segment display can be used for multiple purposes. One of them is to display times, which is exactly what we want to do in our final project. But how can we control them?

The 4-digit display has 12 pins: one pin for each digit (from 0 to 9), one pin for each segment, and a pin for the dot. So, to display anything, we have to send a high signal to the digit we want to set and then just set all pins to high, which we need to represent the character we want to display.

For instance, if we want to display the character of "1" in the fourth digit, we would set pin 4 and pins B and C to high.

To get a better understanding of this, take a look at the following diagram:

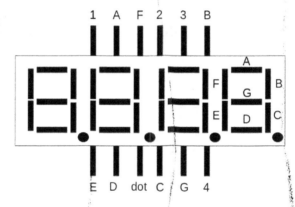

Figure 5.10 – A 7-segment display pinout

From the preceding diagram, you can see that pins 1 to 4 are being used to select the digit.

The 7-segment A-G pins are being used to control the segments and the **dot** pin is being used to set the dot.

So, having to control 12 pins is kind of hard, as then we would only have 2 digital pins left when controlling the display. That is why we use another device to control the display: a MAX7219. The next section explains how this is done.

Using a MAX7219

The MAX7219 (and Max7221) is a *serially interfaced*, 8-digit LED *display driver*. In short, we can control this chip using only four wires, which controls up to eight 7-segment digits.

To send data to that chip, we simply need to drive the load pin low, send 1 byte of data containing the register to set and 1 byte of data to set the segments and the dot. Then, we drive the load pin high and the 16 bits that have been written are processed. The chip will then decode the data and set all of the output pins. The following diagram is a pinout of the chip for reference:

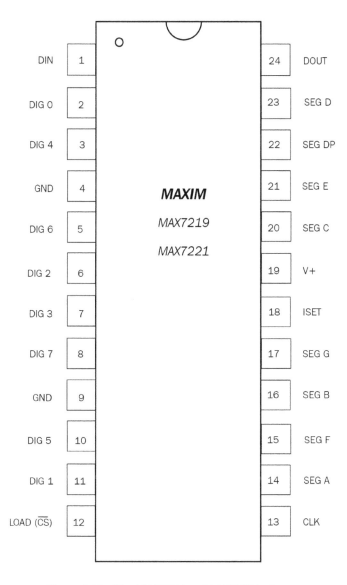

Figure 5.11 – The MAX7219 and MAX7221 pinout

These chips are commonly used in 8x8 LED matrices. So ,if you have such a device, you could carefully remove the chip using a tweezer. *Removing the chip could cause permanent damage to the chip!* These chips are also freely available in most shops for microcontroller components. Before we start to write a library for this chip, let's first assemble our circuit. To do this, perform the following steps:

1. Take a second breadboard; a half-size one is sufficient.

2. Place the MAX7219 on to the breadboard. The *CLK* pin should sit in *E1*, and the *LOAD* pin should sit in *G1*.

3. Place the 7-segment display on to the breadboard. The E pin should sit in D25, and the 1 pin should sit in *F25* or *G25* (depending on which fits better).

4. Connect *DIG0* from the MAX7219 with *Digit1* from the display.

5. Connect *DIG1* from the MAX7219 with *Digit2* from the display.

6. Connect *DIG2* from the MAX7219 with *Digit3* from the display.

7. Connect *DIG3* from the MAX7219 with *Digit4* from the display.

8. Connect *SEGA* from the MAX7219 with *A* from the display.

9. Connect *SEGB* from the MAX7219 with *B* from the display.

10. Connect *SEGC* from the MAX7219 with *C* from the display.

11. Connect *SEGD* from the MAX7219 with *D* from the display.

12. Connect *SEGE* from the MAX7219 with *E* from the display.

13. Connect *SEGF* from the MAX7219 with *F* from the display.

14. Connect *SEGG* from the MAX7219 with *G* from the display.

15. Connect *SEGDP* from the MAX7219 with *DOT* from the display.

16. Connect the *GND* lane from the power bus of both breadboards.

17. Connect the *VCC* lane from the power bus of both breadboards using a jumper wire.

18. Connect *ISET* with VCC from the MAX7219 using a **10,000-Ohm resistor**. This is a hardware solution that controls the brightness of the display.

19. Connect *D13* from the Arduino with *CLK* from the MAX7219.

20. Connect *D6* from the Arduino with *LOAD* from the MAX7219.

21. Connect *D11* from the Arduino with *DIN* from the MAX7219.

22. Connect *D5* from the Arduino with *VCC* from the buzzer.

23. Connect *GND* from the buzzer with *GND* on the power bus.

When all of this is done, the result should now look similar to the following diagram:

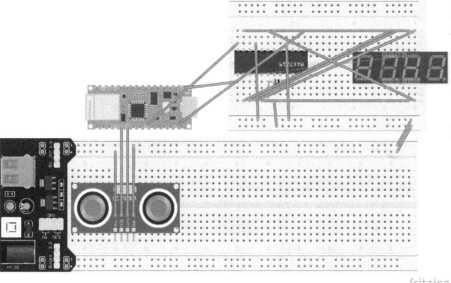

Figure 5.12 – The final circuit

Now, let's better understand how the MAX7219 works by writing a small library that communicates with this chip.

Writing a library to control the MAX7219

We do not only want to learn how to use a MAX7219 in a single project, but we also want to create a library that we can use in all future projects, even beyond the book.

To begin with, we need to create a new folder, called `max7219spi`, inside the `Chapter05` folder. Create two files named `registers.go` and `device.go` inside the newly created folder, and use `MAX7219spi` as the package name. The project structure should look like the following:

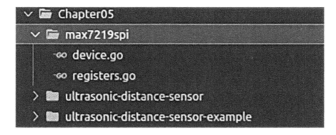

Figure 5.13 – The project structure to control the MAX7219

We are now ready to move ahead and write some code. We will start by implementing the needed registers in the next section.

Registers.go

Inside the `registers.go` file, we place a bunch of constants that represent register addresses. We will explain the constants as soon as we use them in the code:

```
const (
    REG_NOOP byte = 0x00
    REG_DIGIT0 byte = 0x01
    REG_DIGIT1 byte = 0x02
    REG_DIGIT2 byte = 0x03
    REG_DIGIT3 byte = 0x04
    REG_DIGIT4 byte = 0x05
    REG_DIGIT5 byte = 0x06
    REG_DIGIT6 byte = 0x07
    REG_DIGIT7 byte = 0x08
    REG_DECODE_MODE byte = 0x09
    REG_INTENSITY byte = 0x0A
    REG_SCANLIMIT byte = 0x0B
    REG_SHUTDOWN byte = 0x0C
    REG_DISPLAY_TEST byte = 0x0F
)
```

This is it for this file. Further explanations regarding these constants can also be found in the datasheet at `https://datasheets.maximintegrated.com/en/ds/MAX7219-MAX7221.pdf`. Now, we are going to implement the driver.

Device.go

In this file, we can define an interface to implement its methods. The interface will provide a function to write data to the MAX7219, along with some convenience functions to start a display test for our example. To implement it, follow these steps:

1. Define the `Device` interface with all of its functions, as follows:

    ```
    type Device interface {
        WriteCommand(register, data byte)
        Configure()
    ```

```
        StartShutdownMode()
        StopShutdownMode()
        StartDisplayTest()
        StopDisplayTest()
        SetDecodeMode(digitNumber uint8)
        SetScanLimit(digitNumber uint8)
    }
```

2. Then, define a `device` struct that contains an **SPI bus** and a pin that handles the `load`. We will explain these pins in more detail as soon as we use them:

```
    type device struct {
        bus machine.SPI
        load machine.Pin
    }
```

3. Then, define a function, called `NewDevice`, that creates a new instance of `device` and sets the `load` pin along with the SPI bus:

```
    func NewDevice(
            load machine.Pin, bus machine.SPI) Device {
        return &device{
            load: load,
            bus: bus,
        }
    }
```

4. Define a function, called `WriteCommand`, that takes 2 bytes as its parameter. The first byte is `register` and the second one is the `data` to set. For the register, we use the constants of the `registers.go` file. We write data to the MAX7219 by pulling the `load` pin to low. Next, we write the `register` byte, then the `payload`, and then we pull the load `pin` to `high`. Pulling the `load` pin to `high` triggers the MAX7219 to load and process the data:

```
    func (driver *device) WriteCommand(
            register, data byte) {
        driver.load.Low()
        driver.writeByte(register)
```

```
driver.writeByte(data)
driver.load.High()
}
```

5. Define a function, called `Configure`, that sets the `load` pin as the output:

```
func (driver *device) Configure() {
    outPutConfig := machine.PinConfig{
        Mode: machine.PinOutput,
    }
    driver.load.Configure(outPutConfig)
}
```

6. Define a function, called `SetScanLimit`, that tells the MAX7219 how many digits we are going to use in our program. Digits in the MAX7219 start from 0, so we subtract 1 from our digit number, as follows:

```
func (driver *device) SetScanLimit(digitNumber uint8) {
    driver.WriteCommand(REG_SCANLIMIT, byte(
        digitNumber-1))
}
```

7. Next, define a function, called `SetDecodeMode`, that tells the MAX7219 how many digits should be decoded. The decode mode is going to help us later, as it translates our input into the matching output for the 7-segment display in order to display a character. The MAX7219 has a predefined set of characters for this purpose:

```
func (driver *device) SetDecodeMode(digitNumber uint8) {
```

8. Switch over to the `digitNumber` input; if we only use one digit, we tell the MAX7219 to only *decode the first digit*:

```
switch digitNumber {
case 1: driver.WriteCommand(REG_DECODE_MODE, 0x01)
```

If we use two, three, or four digits, we tell the MAX7219 to *decode the first four digits*:

```
case 2, 3, 4: driver.WriteCommand(
    REG_DECODE_MODE, 0x0F)
```

9. Decode all of the digits:

```
case 8: driver.WriteCommand(REG_DECODE_MODE, 0xFF)
```

10. If the input is 0, or greater than 8, we tell the MAX7219 to *decode nothing*:

```
default: driver.WriteCommand(REG_DECODE_MODE, 0x00)
        }
    }
```

We want to be able to activate and deactivate the shutdown mode. We do this by writing a byte to the REG_SHUTDOWN register. This looks like the following snippet:

```
func (driver *device) StartShutdownMode() {
    driver.WriteCommand(REG_SHUTDOWN, 0x00)
}
func (driver *device) StopShutdownMode() {
driver.WriteCommand(REG_SHUTDOWN, 0x01)
}
```

11. Now, we want to be able to start and stop the display test mode. The *display test activates all of the connected LEDs*. This looks like the following:

```
func (driver *device) StartDisplayTest() {
    driver.WriteCommand(REG_DISPLAY_TEST, 0x01)
}
func (driver *device) StopDisplayTest() {
    driver.WriteCommand(REG_DISPLAY_TEST, 0x00)
}
```

12. Define a function, called writeByte, that takes a byte and writes it to the MAX7219. Here, we make use of the SPI interface. First, the SPI implementation internally pulls down the clock pin, then it takes each bit of the byte, and sets the data pin to low for a 0 and to high for a 1. After the data line bit has been set, it pulls up the clock pin:

```
func (driver *device) writeByte(data byte) {
    driver.bus.Transfer(data)
}
```

This is everything we need for the MAX7219.

In the next section, we will create a small abstraction layer above this device. Our abstraction layer will implement the specifics of 7-segment displays. We have implemented the MAX7219 package in a very generic way, and that was done on purpose so that we can build abstraction layers for 7-segment displays and 8x8 LED matrices based on this package.

Writing a library to control the hs42561k display

This library makes use of the MAX7219 library to set it up for 7-segment display use and provides a convenience function to set a character to a specific digit. We start by creating a new folder, called `hs42561k`, inside the `Chapter05` folder and create two files named `constants.go` and `device.go`. Then, name the package `hs42561k`. The project structure should look similar to the following:

Figure 5.14 – The project structure to control the hs42561k display

We start with the `constants.go` file. This file is going to hold some constants and a convenience function that returns a string for a character. To do this, perform the following steps:

1. Add constants for all of the characters. The values are taken from the MAX7219 datasheet. If we use these values, the MAX7219 is going to set the correct pins on the display, thanks to the integrated decoder:

```
const (
    Zero Character = 0
    One Character = 1
    Two Character = 2
    Three Character = 3
    Four Character = 4
    Five Character = 5
    Six Character = 6
    Seven Character = 7
    Eight Character = 8
    Nine Character = 9
    Dash Character = 10
```

```
        E Character = 11
        H Character = 12
        L Character = 13
        P Character = 14
        Blank Character = 15
        Dot Character = 128
    )
```

2. Now, let's add the `Character` struct, which implements the `String` function. The `String` function will come in handy when debugging. We have truncated the list in the example; of course, you might also want to add cases `One` to `Eight`:

```
    type Character byte

    func (char Character) String() string {
        switch char {
        case Zero:
            return "0"
            [...]
        case Nine:
            return "9"
        case Dash:
            return "-"
        case E:
            return "E"
        case H:
            return "H"
        case L:
            return "L"
        case P:
            return "P"
        case Blank:
            return ""
        case Dot:
            return "."
        }
        return ""
    }
```

This is everything we need in the `constants.go` file.

Now, let's implement the `device.go` file by following these steps:

1. Add an interface named `Device` with a `Configure` function and a
    `SetDigit` function:

    ```
    type Device interface {
        Configure()
        SetDigit(digit byte, character Character) error
    }
    ```

2. Add a `struct`, called `device`, which holds the number of digits we want to
    control and a reference to the MAX7219 device:

    ```
    type device struct {
        digitNumber uint8
        displayDevice MAX7219spi.Device
    }
    ```

3. Add a function, called `NewDevice`, that returns a `Device` instance:

    ```
    func NewDevice(displayDevice MAX7219spi.Device,
        digitNumber uint8) Device {
            return &device{
                displayDevice: displayDevice,
                digitNumber: digitNumber,
            }
    }
    ```

4. Add a function named `Configure`. The `Configure` function is used to initialize
    the display driver. It does this by setting the correct `decode` mode and `scan`
    `limit` functions and stopping the shutdown mode to bring the display into
    operational mode, which is implemented in the following snippet:

    ```
    func (device *device) Configure() {
        device.displayDevice.StopDisplayTest()
        device.displayDevice.SetDecodeMode(
            device.digitNumber)
        device.displayDevice.SetScanLimit(
    ```

```
        device.digitNumber)
    device.displayDevice.StopShutdownMode()
```

5. Write `blank` next to each digit, so we can start off with a clean display, just like the following code:

```
for i := 1; i < int(device.digitNumber); i++ {
    device.displayDevice.WriteCommand(byte(i),
        byte(Blank))
    }
}
```

6. Now we can define an *error* for an invalid digit selection:

```
var ErrIllegalDigit = errors.New("Invalid digit
    selected")
```

7. The next step is to define a function, called `SetDigit`, that sets the given character to the given digit:

```
func (device *device) SetDigit(digit byte, character
    Character) error {
```

8. If we have an invalid digit number, we need to validate the `digit` input and return an error. This is because we cannot display values on digits that do not exist:

```
if uint8(digit) > device.digitNumber {
    return ErrIllegalDigit
}
```

9. The last step is to write the `character` to the given `digit`, as shown in the following snippet:

```
device.displayDevice.WriteCommand(
    digit, byte(character))
return nil
}
```

This is the complete logic for the display driver.

Now, let's add a small example project to validate that our code is working as expected. For that purpose, we create a new folder, called `hs42561k-spi-example`, inside the `Chapter05` folder and create a new `main.go` file with an empty `main` function in it. The project structure should look like the following:

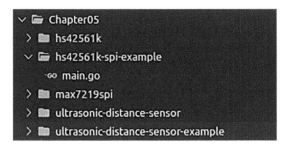

Figure 5.15 – The project structure to validate the code for the hs42561k-spi-example

Now, we can add logic to the new `main.go` file. Follow these steps to set up our example program:

1. First, we add an array of `Character`, which contains all possible characters. Here are the characters that we want to display:

```go
var characters = [17]hs42561k.Character{
    hs42561k.Zero,
    hs42561k.One,
    hs42561k.Two,
    hs42561k.Three,
    hs42561k.Four,
    hs42561k.Five,
    hs42561k.Six,
    hs42561k.Seven,
    hs42561k.Eight,
    hs42561k.Nine,
    hs42561k.Dash,
    hs42561k.E,
    hs42561k.H,
    hs42561k.L,
    hs42561k.P,
    hs42561k.Blank,
    hs42561k.Dot,
}
```

2. Configure the SPI0 interface. SDO is our output pin, and SCK is our clock pin. We send data with the most *significant bit first* at a frequency of 10 MHz. 10 MHz is the maximum frequency that the MAX7219 can handle according to the datasheet:

```
err := machine.SPI0.Configure(machine.SPIConfig{
    SDO: machine.D11,
    SCK: machine.D13,
    LSBFirst: false,
    Frequency: 10000000,
})
```

3. We check whether there was an error and print the error if there was one. This information helps us when debugging:

```
if err != nil {
    println("failed to configure spi:", err.Error())
}
```

4. Initialize the MAX7219 display driver with D6 as the load pin and machine.SPI0 as the SPI bus:

```
displayDriver := max7219spi.NewDevice(
    machine.D6, machine.SPI0)
displayDriver.Configure()
```

5. Now, we need to initialize the display with 4 digits. After this step, the display is ready to be used:

```
display := hs42561k.NewDevice(displayDriver, 4)
display.Configure()
```

6. For each character in characters, set the character to all digits and sleep for half a second. That way, we can test whether we are able to display every possible character on every digit:

```
for {
    for _, character := range characters {
        println("writing", "characterValue:",
            character.String())
        display.SetDigit(4, character)
        display.SetDigit(3, character)
```

```
            display.SetDigit(2, character)
            display.SetDigit(1, character)
            time.Sleep(500 * time.Millisecond)
        }
    }
}
```

This is the complete example program. We can go ahead and flash the program using the following command:

```
tinygo flash --target=arduino-nano33 Chapter05/hs42561k-spi-
example/main.go
```

If everything went as expected, the display should now start to print each possible character.

We have now learned how to control a 7-segment display, learned about the MAX7219 display driver, wrote a library for the display driver and the display, and also wrote an example program. In the next section, we are going to use these libraries and the ultrasonic distance sensor to build our final project of this chapter.

Putting it all together

In our final project of this chapter, we are going to make use of everything we have learned in the preceding sections. We are going to use the ultrasonic distance sensor to recognize a hand movement in close proximity to the sensor. We are using the 7-segment display to count down from 20 to 0 and we are going to use a buzzer, to provide an additional signal, for the timer start and the timer end. In Germany, it is officially recommended that we wash our hands for at least 20 seconds, which is why we will also add a timer for 20 seconds. Putting all of this together, we will create a touchless handwash timer.

Before we start to write the code to control the hand wash timer, we need to add a buzzer. We can add this by following these steps:

1. Put the *GND* pin of the buzzer in *D53* and the *VCC* pin of the buzzer into D54. If that is too close together for your buzzer's pins, just put the buzzer in and wire the following two wires accordingly.

2. Connect the *GND* pin of the Arduino with *A53* on the breadboard using a jumper wire.

3. Connect the *D5* pin of the Arduino with *A54* on the breadboard using a jumper wire.

The circuit should look similar to the following diagram:

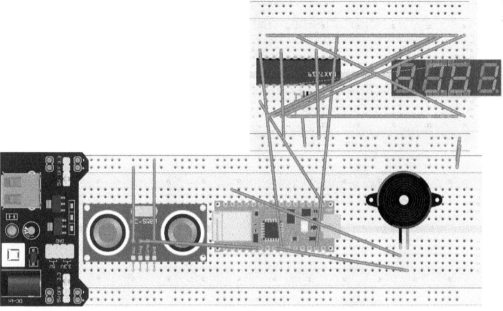

Figure 5.16 – The touchless handwash timer circuit

Now that we have set up the circuit, we can go ahead and write the logic. We start by creating a new folder, called `touchless-handwash-timer`, inside the `Chapter05` folder. Then, we create a new `main.go` file with an empty `main` function. The project structure should look like the following:

Figure 5.17 – The project structure for the handwash timer

Now, follow these steps inside the `main` function to implement the logic for the touchless handwash timer:

1. The first step is to initialize the `SPI0` interface, as follows:

    ```
    err := machine.SPI0.Configure(machine.SPIConfig{
            SDO: machine.D11,
            SCK: machine.D13,
            LSBFirst: false,
            Frequency: 10000000,
    })
    ```

2. If an error occurs, we print it. Doing so enables us to debug the program by monitoring the output of the serial port:

    ```
    if err != nil {
        println("failed to configure spi:", err.Error())
    }
    ```

3. Now we want to initialize the `display`:

    ```
    displayDriver := max7219spi.NewDevice(
        machine.D6, machine.SPI0)
    displayDriver.Configure()
    display := hs42561k.NewDevice(displayDriver, 4)
    display.Configure()
    ```

4. After that is done, we can go on and initialize the `distanceSensor`:

    ```
    distanceSensor := hcsr04.NewHCSR04(
        machine.D2, machine.D3, 60)
    distanceSensor.Configure()
    ```

5. Now, we initialize the `buzzer`. If you skipped *Chapter 4, Building a Plant Watering System*, simply import the buzzer package from `https://github.com/ PacktPublishing/Creative-DIY-Microcontroller-Projects- with-TinyGo-and-WebAssembly/tree/master/Chapter04/buzzer`:

    ```
    buzzer := buzzer.NewBuzzer(machine.D5)
    buzzer.Configure()
    ```

6. Now, we get and print the `currentDistance`. Printing the distance helps us to debug the program if any problems occur later on. This looks like the following:

```
for {
    currentDistance := distanceSensor.GetDistance()
    println("current distance:", currentDistance)
```

7. If the `currentDistance` is between 12 and 25 centimeters, activate the timer. This is shown with the following snippet:

```
if currentDistance >= 12 && currentDistance <= 25 {
    println("timer activated")
    handleTimer(display, displayDriver, buzzer)
}
```

8. Now we have to sleep for `100` milliseconds. We do this to prevent the echoes from overlapping:

```
time.Sleep(100 * time.Millisecond)
}
```

9. The last function is the `handleTimer` function, which takes `display`, `displayDriver`, and `buzzer` as parameters:

```
func handleTimer(display hs42561k.Device,
    displayDriver max7219spi.Device, buzzer
        buzzer.Buzzer) {
```

10. First, we make sure that `display` is in operational mode:

```
display.Configure()
```

11. Now we let the `buzzer` beep two times to indicate that the timer has started:

```
buzzer.Beep(100*time.Millisecond, 2)
```

12. Then, we count from 20 to 0. This represents the 20 seconds that our timer is running, as follows:

```
for i := 20; i > 0; i-- {
    println("counting:", i)
```

13. If we have more than 10 seconds left, we need to set the third `digit`. Because we need to set more than one digit, we are going to set digit 3 and 4, which looks like the following:

```
if i >= 10 {
    display.SetDigit(3, hs42561k.Character(i/10))
```

Additionally, we need to handle all numbers that have a trailing 0. And this looks similar to the following snippet:

```
if i%10 == 0 {
    display.SetDigit(4, hs42561k.Character(0))
} else {
    display.SetDigit(4, hs42561k.Character(i-10))
}
```

Now, we need to handle all numbers that are smaller than 10. This is implemented in the following snippet:

```
} else {
    display.SetDigit(3, hs42561k.Blank)
    display.SetDigit(4, hs42561k.Character(i))
}
time.Sleep(time.Second)
}
```

14. After the timer runs out, we reset both used digits by setting them to `blank`:

```
display.SetDigit(3, hs42561k.Blank)
display.SetDigit(4, hs42561k.Blank)
```

Let the `buzzer` beep for half a second to indicate that the timer has finished:

```
buzzer.Beep(500*time.Millisecond, 1)
```

15. Put the display driver into shutdown mode:

```
displayDriver.StartShutdownMode()
```

This is all the code we need. Now, try the code by flashing it onto the Arduino using the following command:

```
tinygo flash –target=arduino-nano33 Chapter05/touchless-
handwash-timer/main.go
```

We have successfully built and flashed the program. Now it is time to try it out.

So, we combined all the components that we built throughout this chapter into this final project, and we used the components to recognize the movement from a certain distance in front of the sensor to start a timer. This was the final project of this chapter.

Summary

In this chapter, we have learned about the technical specifications of the Arduino Nano 33 IoT and how to calculate the distance between an object and an ultrasonic distance sensor. Additionally, we learned how the sensor works internally and wrote a library for it. We also learned that unit testing is supported in TinyGo and wrote some tests for the ultrasonic distance sensor library. Then, we learned how to use a MAX7219 serial interfaced display driver to control a 7-segment display, and we wrote a library for the MAX7219 and the 7-segment display. At the end of this chapter, we put all of the drivers into a single project and only had to add a small amount of control logic to build a touchless handwash timer.

In the next chapter, we are going to learn how to use 16x02 LCD and ST7735 TFT displays.

Questions

1. Is it possible to draw 5V output from the Arduino Nano 33 IoT?

2. Why do we divide `pulseLength` by 2 when calculating the distance to an object?

3. Change the code so that the handwash timer counts from 120 to 0. Use three digits to display the remaining seconds.

6

Building Displays for Communication using I2C and SPI Interfaces

In the previous chapter, we learned how to display data using a 7-segment display, how a MAX7219 chip works, how ultrasonic distance sensors work, and how to write a library for all this. We used the SPI interface to do so.

After working through this chapter, we will know how to use different types of displays and which displays use different interfaces for communication. We are going to learn how the I2C interface works by using a display that we can connect using an I2C bus. With that covered, we are going to learn how to read and interpret user input. After that, we are going to learn how to draw shapes and texts on displays. Finally, we are going to learn how to build a game that can run on a microcontroller. With this knowledge, we will be able to understand the overall concept of using various displays for communication.

In this chapter, we're going to cover the following main topics:

- Exploring the TinyGo drivers
- Displaying text on a 16x2 LCD display
- Displaying user input on the display
- Building a CLI
- Displaying a simple game

Technical requirements

We are going to need the following components for this project:

- An Arduino Nano 33 IoT
- HD44780 1602 LCD display bundled with an I2C interface
- ST7735 display
- 1 x breadboard
- 1 x 10k Ohm resistor
- 1 x 4-pinned button
- Jumper wires

You can find the code for this chapter on GitHub: `https://github.com/PacktPublishing/Creative-DIY-Microcontroller-Projects-with-TinyGo-and-WebAssembly/tree/master/Chapter06`

The Code in Action video for the chapter can be found here: `https://bit.ly/2Qo8Jji`

Exploring the TinyGo drivers

In *Chapter 3*, *Building a Safety Lock Using a Keypad*, we learned about the TinyGo drivers repository. Let's have a brief look at how to find drivers and examples in this repository.

When you're planning a new project, it is always good to check if the drivers repository has drivers for the devices you plan to use. It will speed up your project and make it easier to implement.

The drivers repository is split into two parts:

- The drivers
- Examples

The drivers directly reside in the root of the repository. All the examples are inside an example folder.

We want to use an hd44780 LCD display with an I2C interface in our example, so let's check if we can find it inside the drivers repository. Refer to the following screenshot:

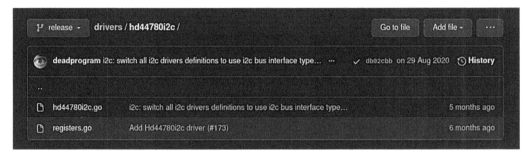

Figure 6.1 – An hd44780i2c driver

As we can see, the package is named after the device and the interface (**I2C**) it uses. Sometimes, a driver package provides more than one interface to use in one package. Most of the drivers omit the additional interface in the name.

To find example code that shows how to use a package, navigate to the `examples` folder and look for a folder that has exactly the same name as the driver package. The following screenshot shows the example code for the **hd47780i2c** driver:

Figure 6.2 – hd44780i2c driver example

Now that we now know that there is a driver for the display we want to use and where to find example code for that driver, let's move on and use that driver.

Displaying text on an HD44780 16x2 LCD display

The HD44780 16x2 LCD is cheap and easy to use. If we only want to display text, this type of display can do just that and is the device of choice. It has 16 pins, which is too many, if we want to combine it with more devices in a project. That is why it is a pretty common practice to use an I2C display driver to control the display. This is a concept similar to using a MAX7219 to drive a 7-segment display, as we did in the previous chapter.

The HD44780 16x2 display can be obtained in a bundle with an I2C driver soldered to it, or it can come without an I2C driver. The display can come in different color configurations, pertaining to background and text color. They typically look similar to the one shown in the following image:

Figure 6.3 – HD44780 front

When the display comes with an I2C driver, it is usually an LCM1602 IIC, which provides four ports:

- GND
- VCC
- SDA
- SCL

So, when using the LCM1602, we only need to connect *GND* and *VCC* to the power bus; the remaining two wires are used for *SDA* and *SCL*. The LCM1602 IIC has a potentiometer on the board, which can be used to adjust the contrast of the display. The following image shows such an LCM1602 IIC, which has been soldered to the back of an HD44780:

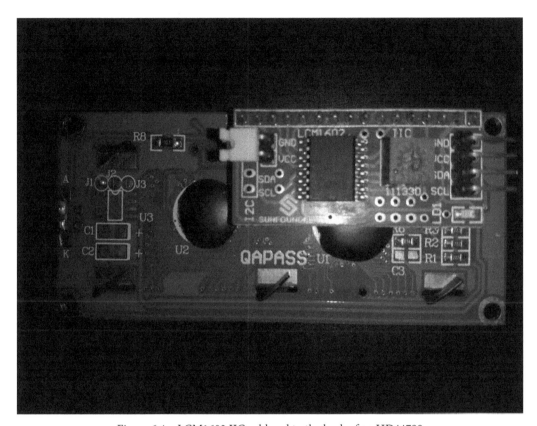

Figure 6.4 – LCM1602 IIC soldered to the back of an HD44780

> **Note**
> Most HD47780 displays operate at 5V, but some only need 3.3V. So, check the datasheet of your display carefully to prevent possible damage!

We now have a brief understanding of the HD44780 and that we can utilize an LCM1602 IIC to save some pins. Now, let's move on and build the circuit.

Building the circuit

Before we can show anything on the display, we need to build the circuit. Just follow these steps to do so:

1. Make sure that the jumper of the power supply sits on 5V. Double-check if you might have a 3.3V display and if so, set the jumper to 3.3V.

2. Connect the *GND* pin of the display to the *GND* lane on the power bus.

3. Connect the *VCC* pin of the display to the *VCC* lane on the power bus.

4. Connect *A14* to the breadboard (*GND*) with the *GND* lane on the power bus.

5. Connect *A9* to the breadboard (*SCL*) with the *SCL* pin of the display.

6. Connect *A8* to the breadboard (*SDA*) with the *SDA* pin of the display.

The circuit should now look similar to the following:

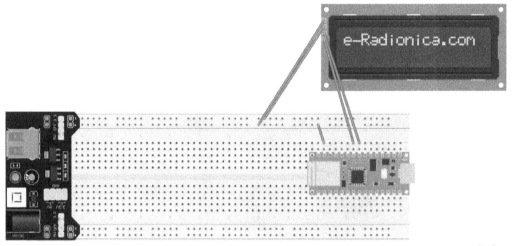

Figure 6.5 – 16x02 I2C display circuit (image taken from Fritzing)

> **Note**
>
> 16x02 I2C LCD Fritzing parts have been taken from the following link:
>
> `https://github.com/e-radionicacom/e-radionica.com-Fritzing-Library-parts-`.

That was everything we needed to set up regarding our hardware devices. However, before we start writing the code, we need to understand I2C.

Understanding I2C

I2C is a synchronous two-wired serial bus, where the data wire being used is bidirectional. Sometimes, **I2C** is also called a **Two-Wire Interface** (**TWI**). One wire is used to provide a **clock**, while the other wire is used to **transmit data**.

The I2C bus allows multiple devices to communicate on the same bus. Unlike the **Peripheral Interface** (**SPI**)) bus, the I2C bus does not need a **chip select** (**CS**) pin; instead, it just includes the address of the receiving device in the message.

An I2C message contains the following parts:

- **Start condition**: The start condition signals that a new message is being sent.

- **Address frame**: The address frame contains the address of the device that should receive the message.

- **Read/Write bit**: This bit is used to signal whether data is being sent from the controller to the device, or if data is being requested from the device.

- **ACK/NACK bit**: The receiving device tells the sender if the previous frame has been received successfully.

- **Data frame**: A single message can contain *1* to *n* DataFrames of 8 bits each.

- **Stop condition**: The stop condition signals that the message has been completely sent.

The following image visualizes a message with 16 bits of data:

Figure 6.6 – I2C message

> **Note**
>
> If you do not know the address of the device you want to use, you can make use of the ACK bit by iterating all possible addresses and checking if the device sends an ACK on an address. If that is the case, you've found the address.

Now that we have a brief understanding of what I2C is and how it works, we can write our first program using I2C to control the display.

Writing the code

We will start by creating a new folder named `Chapter06` inside our project. Inside the `Chapter06` folder, create a new folder named `hd44780-text-display` and create a new `main.go` file with an empty `main` function inside it. The project structure should now look as follows:

Figure 6.7 – Project structure

Now, follow these steps to display the first piece of text:

1. Import the driver:

    ```
    "tinygo.org/x/drivers/hd44780i2c"
    ```

2. Inside the `main` function, configure the I2C interface and set the clock's frequency to `400KHz`:

    ```
    machine.I2C0.Configure(machine.I2CConfig{
        Frequency: machine.TWI_FREQ_400KHZ,
    })
    ```

3. Create a new instance of `hd44780i2c` and pass the `I2C` interface, as well as `address`, as a parameter. Most LCM1602 IICs should listen on the `0x27` address, but some modules listen on `0x3F`:

    ```
    lcd := hd44780i2c.New(machine.I2C0, 0x27)
    ```

4. Configure the display by setting columns (`Width`) and rows (`Height`). We need to do this as this driver also supports 20x4 and other types of displays:

    ```
    lcd.Configure(hd44780i2c.Config{
        Width: 16,
        Height: 2,
    })
    ```

5. Print the text. `\n` is being interpreted by the driver, and all characters followed by `\n` are being written to the next row. We can do this with the following code:

    ```
    lcd.Print([]byte(" Hello World \n LCD 16x02"))
    ```

6. Now, let's test the code by flashing it. Use the following command:

```
tinygo flash --target=arduino-nano33 Chapter6/hd44780-
text-display/main.go
```

You should now see the text being printed on the screen.

Let's look at what happens when we try to print more than 16x2 characters onto the screen. To do so, just add the following snippet to the end of our `main` function:

```
time.Sleep(5 * time.Second)
lcd.Print([]byte("We just print more text, to see what
    happens, when we overflow the 16x2 character limit"))
```

Now, flash the program again and look at the result. What we can observe is that after reaching the 32nd character, the cursor jumps to position x = 0 and y = 0 again and continues to print from there. However, we want to print more than 32 characters on the display, and we want to be able to read all of them. To do so, we must create a small animation. Perform the following steps:

1. At the end of the `main` function, sleep for 5 seconds and call the `animation` function and pass `lcd` as a parameter, as shown in the following code snippet:

    ```
    time.Sleep(5 * time.Second)
    animation(lcd)
    ```

2. We need to define the `animation` function, which takes `lcd` as a parameter:

    ```
    func animation(lcd hd44780i2c.Device) {
    ```

3. Now, we need to define the text we want to print:

    ```
    text := []byte(" Hello World \n Sent by \n Arduino
        Nano \n 33 IoT \n powered by \n TinyGo")
    ```

4. We must clear the display to remove everything we printed previously. This also resets the cursor to the first position (0,0):

    ```
    lcd.ClearDisplay()
    ```

5. Now, let's print a single character. We need to do some type conversions here as the display driver only accepts `[]byte` as a parameter. For this, refer to the following code:

```
for {
    for i := range text {
        lcd.Print([]byte(string(text[i])))
        time.Sleep(150 * time.Millisecond)
    }
```

6. When the message has been completely written onto the display, we sleep for 2 seconds and clear the display again. This enables a clean start for the next iteration:

```
    time.Sleep(2 * time.Second)
    lcd.ClearDisplay()
    }
}
```

Now, flash the updated program again. The characters should now nicely appear one after the other.

Now that we understand how to use the display driver to print hardcoded texts and how to create a simple animation, let's display some dynamically received texts.

Displaying user input on the display

In this section, we are going to print the input of a user onto the display. The input is being sent from the computer to the microcontroller using **serial (UART)**, which will then print it onto the display.

In *Chapter 2, Building a Traffic Lights Control System*, we learned how to use UART to send messages to the computer, and observed them using PuTTY. Now, we are going to use this interface bidirectionally. For this project, we are using the same hardware setup that we used in the previous section, which means we can directly dive into the code.

Start by creating a new folder named hd44780-user-input inside the Chapter06 folder. Then, inside this newly created folder, add a new main.go file with an empty main() function inside it. The project's structure should now look similar to the following:

Figure 6.8 – Project structure

Follow these steps to implement the program:

1. Save the hex value for `carriageReturn` as a constant. Later, we will be checking if a received byte equals this `carriageReturn` value:

```
const carriageReturn = 0x0D
```

2. Save the `uart` interface in a variable so that we don't have to type `machine.UART0` every time:

```
var (
    uart = machine.UART0
)
```

3. Inside the `main` function, start by initializing the display driver:

```
machine.I2C0.Configure(machine.I2CConfig{
    Frequency: machine.TWI_FREQ_400KHZ,
})
lcd := hd44780i2c.New(machine.I2C0, 0x27) // some
                        // modules have address 0x3F
err := lcd.Configure(hd44780i2c.Config{
        Width: 16, // required
        Height: 2, // required
        CursorOn: false,
        CursorBlink: false,
})
if err != nil {
    println("failed to configure display")
}
```

4. Let the user know that we can type something and then print it on the display:

```
lcd.Print([]byte(" Type to print "))
```

5. We want to clear the display as soon as the first input has been received. That is why we save this state:

```
hadInput := false
```

6. If no data resides in the buffer, we don't want to do anything. Incoming data is internally buffered by TinyGo using a ring buffer:

```
for {
    if uart.Buffered() == 0 {
        continue
    }
}
```

7. If we encounter the very first input, we must clear the display and save the state that we had input previously:

```
if !hadInput {
    hadInput = true
    lcd.ClearDisplay()
}
```

8. Next, we read one byte from the buffer and log any possible errors:

```
data, err := uart.ReadByte()
if err != nil {
    println(err.Error())
}
```

9. If a `carriageReturn` is being received, such as because the user pressed the *Enter* key, we also want to print in a new line. We print that character on the display, as well as to `uart`, so that the output in PuTTY and the output on the display behave similarly:

```
if data == carriageReturn {
    lcd.Print([]byte("\n"))
    uart.Write([]byte("\r\n"))
    continue
}
```

10. The last step is to simply print the data onto both outputs:

```
lcd.Print([]byte{data})
uart.WriteByte(data)
}
}
```

Now, we can receive data from a computer that is connected to the microcontroller and print it onto the display. Try it out by flashing the program to the microcontroller by using the following command:

```
tinygo  flash -target=arduino-nano33 Chapter06/hd44780-user-
input/main.go
```

Now, start PuTTy, connect to the microcontroller profile, and start typing to check if the program runs correctly. If everything works correctly, PuTTY should also print what you have written, similar to what's shown in the following screenshot:

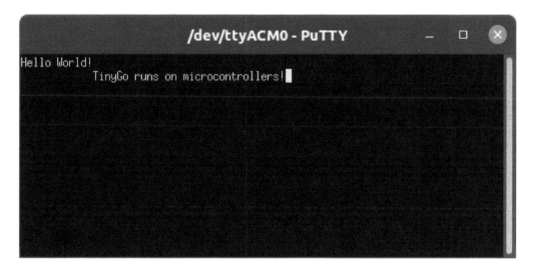

Figure 6.9 – PuTTY output

The UART interface is a serial interface, which means it can be also used to send and receive data between two microcontrollers. On the Arduino Nano 33 IoT, the **transmit (TX)** pin is being used to send the data and the **receive (RX)** pin is being used to receive data.

In this section, we learned how to read and interpret single bytes from the UART interface, as well as how to manually send data back to the UART interface, without using the print() or println() functions. We'll use this knowledge in the next section to learn how to interpret longer strings of data.

Building a CLI

In this section, we are going to parse the input from a user and compare the input with predefined commands. These commands will then be executed by the microcontroller. For this project, we are going to use the same hardware setup that we used in the previous one.

We will start by creating a new folder named hd44780-cli inside the Chapter06 folder. Then, we must create a main.go file with an empty main function inside it. The project's structure should now look similar to the following:

Figure 6.10 – Project structure

Now that the project structure has been set up, we can implement the logic. To do so, follow these steps:

1. Above the main function, start by defining some constants. commandConstant represents the command that needs to be sent to the microcontroller. We will use these constants ahead in this code and compare them with the user input to determine whether a CLI command has been entered:

```
const (
    carriageReturn = 0x0D
    homeCommand = "#home"
    clearCommand = "#clear"
)
```

2. Save the UART interface in a variable. We could also always write `machine.UART0` instead, but by doing it this way, we improve the readability:

```
var (
    uart = machine.UART0
)
```

3. Inside the `main` function, we initialize the display, as follows:

```
machine.I2C0.Configure(machine.I2CConfig{
    Frequency: machine.TWI_FREQ_400KHZ,
})
lcd := hd44780i2c.New(machine.I2C0, 0x27)
err := lcd.Configure(hd44780i2c.Config{
        Width: 16,
        Height: 2,
        CursorOn: false,
        CursorBlink: false,
})
if err != nil {
println("failed to configure display")
}
```

4. Now, let's call the `homeScreen` function (we are going to explain what this function does when we implement it later):

```
homeScreen(lcd)
```

5. Next, define a `commandBuffer`. That is a simple string where we store the parts of a command:

```
var commandBuffer string
```

6. `commandIndex` is being used to count the characters inside `commandBuffer`. If the index is greater than the length of the longest command, then we know that we can reset the buffer:

```
var commandIndex uint8
```

7. We will be using the `commandStart` boolean as a signal, so we need to append any subsequent characters to `commandBuffer`:

```
commandStart := false
```

8. Just like in the previous project, we are going to use the `hadInput` flag to clear the screen when the first input is received:

```
hadInput := false
```

9. We don't need to do anything if there are no characters in the internal receive buffer:

```
for {
    if uart.Buffered() == 0 {
        continue
}
```

10. Upon receiving the first input, clear the display. We will explain the `clearDisplay` function when we implement it ahead, after a few steps:

```
if !hadInput {
    hadInput = true
    clearDisplay(lcd)
}
```

11. Then, we read a byte from the buffer, as follows:

```
data, err := uart.ReadByte()
if err != nil {
    println(err.Error())
}
```

12. Check if we received a **pound sign (#)**. This is the indicator that a command is going to follow:

```
if string(data) == "#" {
    commandStart = true
    uart.Write([]byte("\ncommand started\n"))
}
```

13. When we receive the start of the command, we append all subsequent characters to commandBuffer. This is done as follows:

```
if commandStart {
    commandBuffer += string(data)
    commandIndex++
}
```

14. To check if we may have a complete command inside commandBuffer, we must switch over our commandBuffer:

```
switch commandBuffer {
```

15. If the content of commandBuffer equals homeCommand, we execute the homeScreen function and reset the command. We must also write the input data back inside the UART interface:

```
case homeCommand:
    uart.WriteByte(data)
    homeScreen(lcd)
    commandStart = false
    commandIndex = 0
    commandBuffer = ""
    continue
```

16. If the content of commandBuffer equals clearCommand, we must execute the clearDisplay function and reset the command:

```
case clearCommand:
    uart.WriteByte(data)
    clearDisplay(lcd)
    commandStart = false
    commandIndex = 0
    commandBuffer = ""
    continue
}
```

17. If `commandIndex` is greater than the length of our longest command, we must reset the command:

```
if commandIndex > 5 {
    commandStart = false
    commandIndex = 0
    commandBuffer = ""
    uart.Write([]byte("\nresetting command state\n"))
}
```

18. If we receive a `carriageReturn`, we must print a new line:

```
if data == carriageReturn {
    lcd.Print([]byte("\n"))
    uart.Write([]byte("\r\n"))
    continue
}
```

19. Then, we print the received data, as follows:

```
lcd.Print([]byte{data})
uart.WriteByte(data)
}
```

20. Now, define the `homeScreen` function, which is called when the input matches the `homeScreen` command. We must clear the display and print the first input again:

```
func homeScreen(lcd hd44780i2c.Device) {
    println("\nexecuting command homescreen\n")
    clearDisplay(lcd)
    lcd.Print([]byte(" TinyGo UART \n CLI "))
}
```

21. Now, define the `clearDisplay` function, which is called when the input matches the `clearDisplay` command. We just make use of the `ClearDisplay` function of the display here:

```
func clearDisplay(lcd hd44780i2c.Device) {
    println("\nexecuting command cleardisplay\n")
    lcd.ClearDisplay()
}
```

Now, flash the program using the following command:

```
tinygo flash –target=arduino-nano33 Chapter06/hd44780-cli
```

Now, let's try out our program.

Start putty and select the microcontroller profile. Type something and use the #home and #clear commands that we defined in the code. PuTTY's output should now look similar to the following:

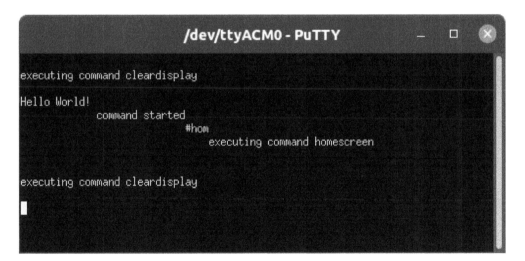

Figure 6.11 – CLI output in PuTTY

With that, we have verified that the program works as intended. Such a system could be used to control a microcontroller using another microcontroller, not just to display something – it could also be used to request sensor readings or trigger other things.

In this section, we learned how to interpret more than a single character of input data at a time, as well as how to set up a simple CLI in order to execute commands that are being sent through UART. In the next section, we are going to gain a deeper understanding of SPI, since we will be using an SPI-driven display in the final project.

Understanding SPI

SPI is a bus system that has a controller and one or many devices. The controller selects a device that should send data to the controller, or that is going to receive data from the controller.

Devices on an SPI bus can also be daisy chained together. A **daisy chain** is a wiring scheme in which you put multiple devices together in a row.

SPI communication between two devices uses the following four pins:

1. **CS**: **ChipSelect** selects which device on the bus should receive or send data.
2. **CLK**: **Clock** sets the frequency of the transfer (DO) and receive (DI) wires.
3. **DO**: **DataOut** or **DigitalOut** transmits data to the receiving device.
4. **DI**: **DataIn** or **DigitalIn** receives data from the controller.

The following diagram shows the one-to-one connection of an SPI controller and an SPI device:

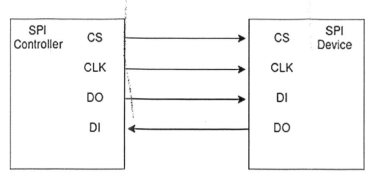

Figure 6.12 – SPI communication

The following diagram shows the SPI connection of one controller and two devices. Here, we are using two CS pins to signal the receiving device. This is the device the controller is talking to:

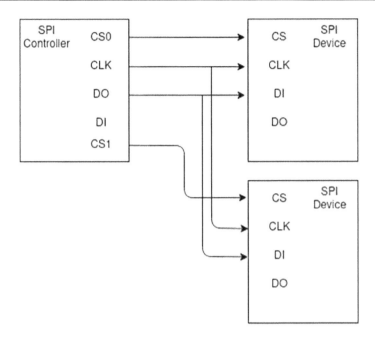

Figure 6.13 – SPI communication between a controller and two devices

The following diagram shows how devices can be daisy chained together. The *DO* pin of the first device is connected to the *DI* pin of the next device, while they share the *CLK* and *CS* wires:

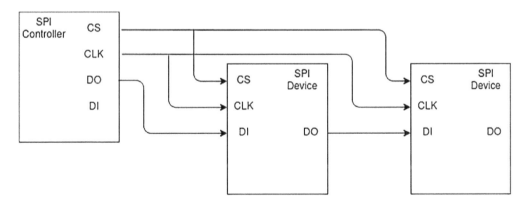

Figure 6.14 – SPI communication with daisy chained devices

Now that we have a better understanding of SPI, let's build a circuit using the ST7735 display.

Displaying a simple game

In this section, we are going to learn how to use another display type using the SPI interface. We need a new type of display since we want to display more than plain text. We will also discover two more TinyGo repositories that provide handy functions for when we're working with displays. The display we are going to use in this section is a 1.8" TFT ST7735 display with a resolution of 160x128 pixels. So, let's have a brief look at the technical specifications of the display.

The ST7735 display provides an SD card slot, which is optional. The display has a color depth of 262K colors on a TFT-LCD module. The SPI interface is being used with the display. To draw something on the display, we need eight pins. We have already used SPI, but we did not have a deeper look at it, since the devices can be arranged on an SPI bus in different ways. So, let's gain a better understanding of how SPI works before we use the display in an example project.

Building the circuit

As in the previous projects, we are going to use an external power supply. We also need a ST7735 display, the Arduino Nano 33 IoT, and some jumper wires. To set everything up correctly, follow these steps:

1. Connect the *GND* lane from the power bus to pin *J50 (GND)* on the breadboard.

2. Connect pin *E31 (LED)* on the breadboard to pin *A53 (D2)* on the breadboard.

3. Connect pin *E32 (SCK)* on the breadboard to pin *J63 (D13)* on the breadboard.

4. Connect pin *E33 (SDA)* on the breadboard to pin *A62 (D11)* on the breadboard.

5. Connect pin *E34 (AO)* on the breadboard to pin *A56 (D5)* on the breadboard.

6. Connect pin *E35 (RESET)* on the breadboard to pin *A57 (D6)* on the breadboard.

7. Connect pin *E36 (CS)* on the breadboard to pin *A58 (D7)* on the breadboard.

8. Connect the *GND* lane from the power bus to pin *E37 (GND)* on the breadboard.

9. Connect the *VCC* lane from the power bus to pin *E38 (VCC)* on the breadboard.

10. Place the ST7735 display so that the *LED* pin sits in *A31* and the *VCC* pin sits in *A37*.

This is everything we need to connect to the display. The setup should now look as follows:

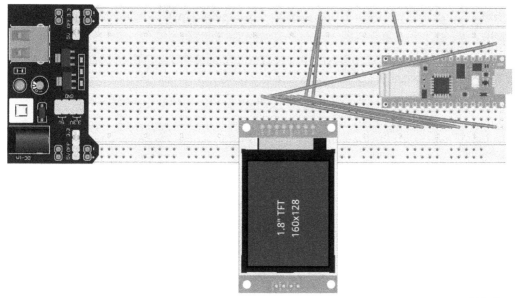

Figure 6.15 – ST7735 circuit

> **Note**
>
> The 1.8″ TFT display Fritzing part is made by vanepp: `https://forum.fritzing.org/u/vanepp`.

Now that we've set up the hardware, let's implement some logic.

Using an ST7735 display

TinyGo provides a driver for ST7735 displays. This means we can use the existing driver. Also, TinyGo provides two additional packages named `TinyFont` and `TinyDraw`, both of which we are going to use. First, let's check out the `TinyDraw` package.

`TinyDraw` is a repository inside the TinyGo organization on GitHub. You can find it at `https://github.com/tinygo-org/tinydraw`.

`TinyDraw` is still in an early state, which means that it has not been optimized for performance or memory usage. However, it provides useful functionality, such as for drawing rectangles, circles, filled rectangles and filled circles, and more. It works with most interface drivers since the APIs of display drivers are nearly (or exactly) the same. Now, let's have a look at the `TinyFont` package before we see it in action.

Just like `TinyDraw`, `TinyFont` is a repository inside the TinyGo organization on GitHub. You can find it at `https://github.com/tinygo-org/tinyfont`.

`TinyFont` provides an API to let you draw text onto displays using fonts that come with the `TinyFont` package. It also allows you to create your own custom font. `TinyFont` also makes use of the fact that most TinyGo display drivers share the same interface.

Now, let's set up a project that uses the ST7735, TinyDraw, and TinyFont. To do so, create a new folder named `st7735` inside the `Chapter06` folder and create a new `main.go` file with an empty `main()` function inside it. The project's structure should now look similar to the following:

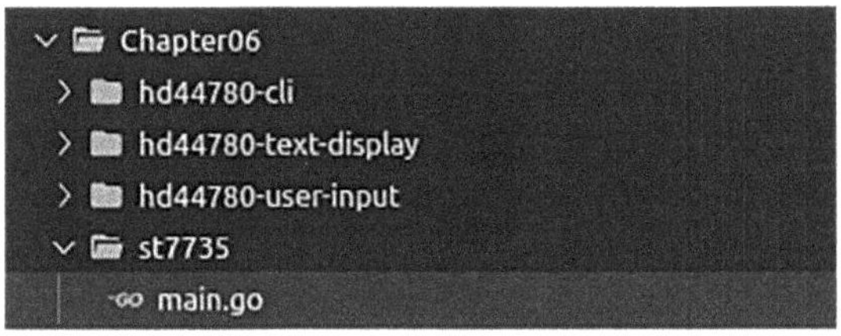

Figure 6.16 – Project structure

Now, let's dive into the code. We will need to import the following packages for this project:

```
"tinygo.org/x/drivers/st7735"
"tinygo.org/x/tinydraw"
"tinygo.org/x/tinyfont"
"tinygo.org/x/tinyfont/freemono"
```

To write our first test program for this display, follow these steps:

1. Above the `main` function, define a set of colors that we are going to use later in the program:

    ```
    var (
        white = color.RGBA{255, 255, 255, 255}
        red = color.RGBA{255, 0, 0, 255}
        blue = color.RGBA{0, 0, 255, 255}
        green = color.RGBA{0, 255, 0, 255}
        black = color.RGBA{0, 0, 0, 255}
    )
    ```

2. Configure the `SPI0` interface with a frequency of 12 MHz. We do not need to pass the pins for SCK and DO as the `Configure` function will use the default SPI pins for this board when no pins are passed:

```
machine.SPI0.Configure(machine.SPIConfig{
    Frequency: 12000000,
})
```

3. Set the required pins for the display:

```
resetPin := machine.D6
dcPin := machine.D5
csPin := machine.D7
backLightPin := machine.D2
```

4. Get a new instance of the `st7735` display:

```
display := st7735.New(machine.SPI0, resetPin, dcPin,
    csPin, backLightPin)
```

5. Call the `Configure` function. This function transmits the bootup sequence to the display. After this call, the display is ready to use:

```
display.Configure(st7735.Config{})
```

6. Get the `width` and `height` attributes from the display:

```
width, height := display.Size()
```

7. Draw four rectangles. Each of them should take up a quarter of the screen and be a different color. The function takes a position on the x-axis, a position on the y-axis, the width and height of the rectangle to draw, as well as the color of the rectangle. We will use this as a test for our display. This is a good test for the display:

```
display.FillRectangle(0, 0, width/2, height/2, white)
display.FillRectangle(width/2, 0, width/2,
    height/2, red)
display.FillRectangle(0, height/2, width/2,
    height/2, green)
display.FillRectangle(width/2, height/2,
    width/2, height/2, blue)
```

8. Before we move on and draw some more advanced graphics, let's test the program by flashing it using the following command:

```
tinygo flash -target=arduino-nano33 Chapter6/st7735/main.
go
```

9. Once the program has been successfully flashed onto the microcontroller, you should see four rectangles. This looks similar to the following:

Figure 6.17 – ST7735 displaying four rectangles

Now, let's draw some more complicated forms using `TinyDraw` and some text using `TinyFont`. To do so, follow these steps:

1. At the end of the `main` function, add a sleep for 3 seconds so that we have a chance to actually see the display test:

```
time.Sleep(3 * time.Second)
```

2. Initialize a counter that will be used to display a count of how many times we have drawn an animation:

```
i := 0
```

3. Fill the screen with `black` to clean the display:

```
for {
    display.FillScreen(black)
```

4. Draw a white rectangle at the lower end of the screen. It should have a height of 32 pixels, which should leave us with 128x128 pixels:

```
tinydraw.FilledRectangle(&display, 0, 0, 128, 32,
    white)
```

5. Since we have assembled our display upside down, we are going to write text rotated:

```
tinyfont.WriteLineRotated(&display,
    &freemono.Bold9pt7b, 110, 145, "TinyDraw", red,
    tinyfont.ROTATION_180)
```

6. At the center of the black square, draw three circles of different sizes and colors on top of each other:

```
tinydraw.FilledCircle(&display, 64, 96, 32, green)
tinydraw.FilledCircle(&display, 64, 96, 24, blue)
tinydraw.FilledCircle(&display, 64, 96, 16, red)
```

7. Now, draw the TinyFont text in green beneath the circles:

```
tinyfont.WriteLineRotated(&display,
    &freemono.Bold9pt7b, 110, 40, "TinyFont", green,
    tinyfont.ROTATION_180)
```

8. Draw the count of how many times the animation will run on the white rectangle:

```
counterText := fmt.Sprintf("Count: %v", i)
tinyfont.WriteLineRotated(&display, &freemono.Bold9pt7b,
123, 2, counterText, black, tinyfont.ROTATION_180)
```

9. We need to sleep for a moment because otherwise, we won't be able to see the result. This is because it will be overridden in the next iteration:

```
time.Sleep(2 * time.Second)
i++
}
```

Now, flash the program again. After the test screen, you should see a result similar to the following:

Figure 6.18 – Result of the test program

In this section, we learned how to draw basic shapes and write text on our display. The next logical step is to write a game that runs on a microcontroller.

Developing a game

In this section, we are going to develop a very simple game that consists of an enemy, represented by a red block, that tries to reach the end of the screen. A green line will represent our home zone, which the red block should not cross. We'll also have a green block that represents the player, as well as a smaller green block that represents a bullet that we can shoot to stop the red block from invading our home zone. We will be adding a button that will act as a trigger and shoot the small green blocks. So, the logical first step is to add the button to our breadboard. To do so, follow these steps:

1. Place the button on the breadboard so that one pin sits in *E23* and the other pins sit in *E25* on one side and *F25* and *F23* on the other side.

2. Connect the *+3V3* output from the Arduino to *J23* on the breadboard.

3. Use a **10K Ohm resistor** to connect the *GND* lane from the power bus to *J25*.

4. Connect *D25* to *A60 (D9)* on the breadboard.

This was everything we needed to add to the circuit. It should now look as follows:

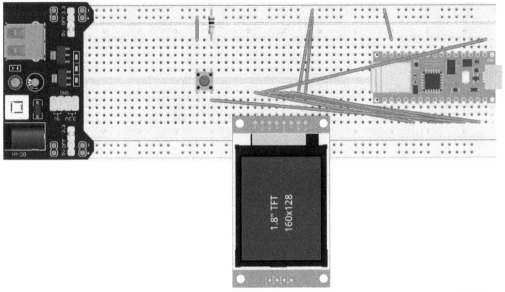

Figure 6.19 – The final circuit with the button

Now, let's create a new folder for the last project in this chapter. Name the folder tinygame and put it inside the Chapter06 folder. Then, create a new main.go file with an empty main() function inside it. The project's structure should now look as follows:

Figure 6.20 – Project structure

To implement the logic, follow these steps:

1. Add a `bool` that holds the `buttonPressed` state. We will define this globally, so we do not need to use channels or something else to pass the state between the goroutines that we are going to use. This is just an easy and convenient way to do this:

   ```
   var buttonPressed bool
   ```

2. Define `enemySize`, `bulletSize`, and the `width` and `height` properties of the game field in pixels:

   ```
   const enemySize = 8
   const bulletSize = 4
   const width = 128
   const height = 160
   ```

3. Add two variables to store our `currentScore` and `highscore`, respectively:

   ```
   var highscore int = 0
   var currentScore int = 0
   ```

4. Define a set of colors that we will use later:

   ```
   var (
       white = color.RGBA{255, 255, 255, 255}
       red = color.RGBA{255, 0, 0, 255}
       blue = color.RGBA{0, 0, 255, 255}
       green = color.RGBA{0, 255, 0, 255}
       black = color.RGBA{0, 0, 0, 255}
   )
   ```

5. Now, we need to move inside the `main` function. Here, assign `buttonPin` and configure it as input:

   ```
   buttonPin := machine.D9
   buttonPin.Configure(machine.PinConfig{Mode:
           machine.PinInput})
   ```

6. Update `highscore` since we are in the startup phase. Here, `highscore` is 0:

   ```
   updateHighscore(0)
   ```

7. Initialize the display, as follows:

```
machine.SPI0.Configure(machine.SPIConfig{
    Frequency: 12000000,
})

resetPin := machine.D6
dcPin := machine.D5
csPin := machine.D7
backLightPin := machine.D2

display := st7735.New(machine.SPI0, resetPin, dcPin,
            csPin, backLightPin)
display.Configure(st7735.Config{})
```

8. Run the `checkButton` function inside a new goroutine so that it is non-blocking. This enables us to update the game loop in the `main` goroutine:

```
go checkButton(buttonPin)
```

9. Loop forever and fill the screen with black to erase everything from the screen after each round of the game:

```
for {
    display.FillScreen(black)
    updateGame(display)
}
```

10. Loop forever and check the button's state. If the button has been pressed, we update the `buttonPressed` state. After each check, we sleep for 20 milliseconds, since we need a blocking call so that the scheduler can work on other goroutines again:

```
func checkButton(buttonPin machine.Pin) {
    for {
        if buttonPin.Get() {
            buttonPressed = true
        }
        time.Sleep(20 * time.Millisecond)
    }
}
```

11. The `updateHighscore` function takes a `score`, checks if this new `score` is greater than `highscore`, and if so, it updates `highscore` and prints `highscore` to the serial:

```
func updateHighscore(score int) {
    if score <= highscore && score != 0 {
        return
    }
    highscore = score
    println(fmt.Sprintf(" TinyInvader Highscore: %d",
        highscore))
}
```

With that, we have implemented a check for button presses, a function to update `highscore`, and also have a main goroutine that starts a new round of the game as soon as it ends. Now, let's implement the actual game logic. To do so, follow these steps:

1. Normally, it would be best to split the update for the game's physics, such as the player's movement, bullets, and the enemy and put it in one part of the logic, and then put the animation in another part of the logic. When developing games for other platforms, these two parts would update independently of each other so that they don't rely on the same framerate. However, to keep things simple, we will have a single game loop that updates the positions as well as drawing to the screen. The `updateGame` function represents the main logic of the game:

```
func updateGame(display st7735.Device) {
```

2. Define some variables that will store the position of the enemy:

```
var enemyPosX, enemyPosY int16
```

3. To prevent the enemy from starting above the game field, we must subtract its size:

```
enemyPosY = height - enemySize
```

4. Next, we need to store the position of the bullet inside a variable:

```
var bulletPosY int16
```

5. We store the state if a shot has been fired in a bool variable:

```
shotFired := false
```

6. We store the state if a new shot can be fired in a bool variable. We initialize it to true as we want the player to be able to fire a shot when the game starts:

```
canFire := true
```

7. The game has just started, so currentScore is 0:

```
currentScore = 0
```

8. If the button has been pressed, we reset the buttonPressed state as we will be handling it. As long as the bullet is still flying inside the game field, we cannot fire again:

```
for {
    if buttonPressed {
        buttonPressed = false
        if canFire {
            shotFired = true
            canFire = false
        }
    }
}
```

9. If a shot has been fired, we update the bullet:

```
if shotFired {
```

10. Here, we update the position and draw it:

```
bulletPosY = updateBullet(display, bulletPosY)
```

If the bullet leaves the game field, we reset the position and reset the shotFired and canFire states. This enables the player to shoot again:

```
if bulletPosY > height {
    shotFired = false
    canFire = true
    bulletPosY = 0
}
```

11. Next, we check that the bullet has collided with the enemy on the horizontal axis. For this, we use a hitbox that is slightly larger than the bullet itself:

```
if enemyPosX >= 54 && enemyPosX <= 64 {
```

12. Now, we check for a collision on the vertical axis. This time, the hitbox is the same size as `bulletSize`. These hitboxes have proven to work pretty well in my tests:

```
if enemyPosY >= bulletPosY && enemyPosY <=
bulletPosY+bulletSize {
```

13. If we hit the enemy, we increment the score:

```
currentScore++
```

14. Now, we must draw a black box over the enemy to let it disappear:

```
display.FillRectangle(enemyPosX-1, enemyPosY,
    enemySize, enemySize, black)
```

15. Reset the enemy's position. This respawns the enemy at its spawn position:

```
enemyPosY = height - enemySize
enemyPosX = 0
```

16. Update `highscore`, like so:

```
updateHighscore(currentScore)
        }
    }
}
```

17. Update and draw the enemy's position:

```
enemyPosX, enemyPosY = updateEnemy(display, enemyPosX,
enemyPosY)
```

18. If the enemy passes our home zone, we lose the game. If this happens, we return, as this lets the loop outside the function run again and start a new game:

```
if enemyPosY < enemySize {
    return
}
```

19. Draw the home zone:

```
display.FillRectangle(0, 4, width, 1, green)
```

20. Draw the player:

```
display.FillRectangle(58, 0, 6, 6, green)
```

21. Sleep for 12 milliseconds. If we do not sleep here, the enemy and the bullet will move too quickly over the screen and will appear to be flickering, which does not look nice. So, we use this little trick to slow it down and reduce the flickering:

```
time.Sleep(12 * time.Millisecond)
    }
}
```

Now that we have implemented the main game logic, we only have to create the logic that updates the bullet and the enemy before we can play the game.

Update the bullet by incrementing its position on the y-axis by 2. Draw a black box behind it so that it does not leave a trail on the display:

```
func updateBullet(display st7735.Device, posY int16)
        int16 {
        display.FillRectangle(58, posY-2, bulletSize, 2,
            black)
        display.FillRectangle(58, posY, bulletSize,
            bulletSize, green)
        return posY + 2
}
```

The last thing we need to do is update the enemy. To do so, follow these last few steps:

1. First, we must define the positions and width of the rectangle we will use to clear out the previous position of the enemy:

```
func updateEnemy(display st7735.Device, posX, posY
        int16) (int16, int16) {
    var clearX, clearY, clearWidth int16
```

2. Now, we must calculate the position where we need to clear out the enemy:

```
clearX = posX - 1
clearY = posY
clearWidth = 1
```

3. If the enemy reaches the left-hand side, we need to completely remove its rectangle as the enemy will spawn on the other side of the screen again:

```
if posX == 0 {
    clearY = posY + enemySize
    clearX = width - enemySize
    clearWidth = enemySize
}
```

4. Now, we must clear out the enemy and draw the enemy in its new position. We must do this to prevent the enemy from leaving trails on the display:

```
display.FillRectangle(clearX, clearY, clearWidth,
enemySize, black)
display.FillRectangle(posX, posY, enemySize, enemySize,
red)
```

5. Update the position of the enemy on the x-axis:

```
posX++
```

6. If the enemy reaches the border of the screen on the x-axis, they also move on the y-axis:

```
if posX > width-enemySize {
    posX = 0
    posY -= enemySize
}
```

7. Return the new position:

```
return posX, posY
}
```

This is all the logic we need for that game. Now, let's play it. Flash the program using the following command:

```
tinygo flash -target=arduino-nano33 Chapter06/tinygame/main.go
```

> **Note**
>
> We do not need to specify a scheduler here because the scheduler is not deactivated by default for atsamd21.

Once you've played the game for a few rounds, you can start thinking about how to extend the game:

- We could add two more buttons so that we can move the player left and right.

- We could make it possible for the player to shoot more than one bullet at a time.

- The enemy's movement could be randomized so that they don't always move from right to left.

- We could add a joystick to control the player's position.

- Multiple enemies could be spawned.

- The enemies could drop different kinds of powerups, which the player could then pick up.

- We could add a buzzer in order to add sounds to the game.

- We could display the high score at the end of each round.

This was the last chapter before we start diving into the world of **IoT** and **WebAssembly**. If you want to test your knowledge, I recommend that you play around with this project. Extend it with more hardware and try to update the game's logic. I already provided some ideas in the preceding list. You could make use of these ideas or think of your own solutions. If you extend the game, I would love to see what and how you did this. So, please feel free to share your games on Twitter using the hashtags `#tinygame`, `#tinygo`, and `#packtbookgame`, and also don't forget to tag me using @Nooby_Games. Of course, you can also share your games on all other social media channels, blogs, and so on. You can also open an issue in this book's GitHub repository to show off your results. That way, I can also playtest your games.

Summary

In this chapter, we learned what the I2C interface is and how to use it. We also learned how to use a 16x02 LCD display, how to display static text, how to display animations, and how to build a little CLI that can receive commands through UART and control the display.

Then, we gained a deeper understanding of the SPI interface and used it to control a 1.8" TFT display. We drew some basic shapes and then used `TinyDraw` to draw circles and rectangles and `TinyFont` to draw text. At this point, we have used all the important interfaces of a microcontroller, so we now have the skills to connect and control any device we need in future projects.

At the end of this chapter, we used the knowledge we'd gained in this chapter to build a simple game that is controlled by one button and is displayed on the 1.8" TFT display.

In the next chapter, we are going to learn how build a **WebAssembly** page using TinyGo, as well as how to use the **Wi-Fi chip** that is built into the Arduino Nano 33 IoT.

Questions

1. How does a device that listens on an I2C bus know that a message is dedicated to that device?

2. How does a device that listens on an SPI bus know that a message is dedicated to that device?

7
Displaying Weather Alerts on the TinyGo Wasm Dashboard

We have learned how to display data using different types of displays that are connected using either the **Inter-Integrated Circuit** (**I2C**) protocol or the **Serial Peripheral Interface** (**SPI**). While doing so, we dived a bit deeper into understanding how SPI works by learning that multiple devices can listen on a SPI bus and that we can daisy-chain devices on the bus. Furthermore, we have built a **command-line interface** (**CLI**) that interprets commands sent over serial and executes functions depending on the input.

After working through this chapter, you will be familiar with using **Message Queuing Telemetry Transport** (**MQTT**), serving a **WebAssembly** (**Wasm**) page through a web server, how to set up a local MQTT broker, and how to use the Wi-Fi functionalities of the Arduino Nano 33 IoT board.

In this chapter, we're going to cover the following main topics:

- Building a weather station
- Sending MQTT messages to a broker
- Introducing Wasm
- Displaying sensor data and weather alerts on a Wasm page

By the end of this chapter, you will know how to utilize an MQTT broker in order to send messages from your microcontroller over Wi-Fi. You will also know how to subscribe to MQTT messages inside a Wasm app and how to display data that is being sent as an MQTT message payload.

Technical requirements

The following software needs to be installed:

- Docker—You can find an installation guide by following this link: `https://docs.docker.com/engine/install/`

We are going to need the following components for this project:

- An Arduino Nano 33 IoT board
- An external power supply (**5 volts (5V)**)
- A BME280 sensor (I2C)
- An ST7735 display
- A breadboard
- Jumper wires

You can find the code for this chapter on GitHub at the following link: `https://github.com/PacktPublishing/Creative-DIY-Microcontroller-Projects-with-TinyGo-and-WebAssembly/tree/master/Chapter07`

The Code in Action video for the chapter can be found here: `https://bit.ly/3dXGe4o`

Building a weather station

We start our journey through the world of **Internet of Things** (**IoT**) and Wasm by building a **weather station**. In our first project of this chapter, we are going to build a program that displays weather data on an **ST7735** display. We are going to build some reusable components that we are going to utilize in the final project of the chapter. We are going to learn how to use a **BME280** sensor that is able to sense air pressure, temperature, and humidity—the elements required for noting a change in the weather. But first, we need to assemble a circuit—so, let's see how that works.

Assembling the circuit

Before we are able to read and display the sensor data, we need to assemble the circuit. We are connecting the BME/BMP280 sensor using the I2C interface, and we are going to connect the ST7735 display using the SPI interface. To do so, perform the following steps:

1. Place the **BME/BMP280** sensor with serial data pin (**SDA**) in *F21*.

2. Connect pin *H21* (SDA) with pin *J56* (SDA) on the breadboard, using a jumper wire.

3. Connect pin *I22* **Serial Clock** (**SCL**) with pin *I55* (SCL) on the breadboard, using a jumper wire.

4. Connect *J23* **Ground** (**GND**) with the *GND* lane on the power bus.

5. Connect *J24* (VIN) with the **Voltage Common Collector** (**VCC**) lane on the power bus.

6. Place the display with the **light-emitting diode** (**LED**) pin in pin *A31* on the breadboard.

7. Connect *E31* (LED) with pin *A53* (D2) on the breadboard, using a jumper wire.

8. Connect *E32* (SCK) with pin *A54* (D13) on the breadboard, using a jumper wire.

9. Connect *E33* (SDA) with pin *A62* (D11) on the breadboard, using a jumper wire.

10. Connect *E34* **Analog pin** (**AO**) with pin *A56* on the breadboard, using a jumper wire.

11. Connect *E35* (AO) with pin *A57* (D5) on the breadboard, using a jumper wire.

12. Connect *E36* (RESET) with pin *A58* (D6) on the breadboard, using a jumper wire.

13. Connect *E37* **chip select** (**CS**) with pin *A59* (D7) on the breadboard, using a jumper wire.

14. Connect *E37* (GND) with the *GND* lane on the power bus.

15. Connect *E38* (VCC) with the *VCC* lane on the power bus.

16. Connect *J51* (GND) with the *GND* lane on the power bus.

The circuit should now look similar to this:

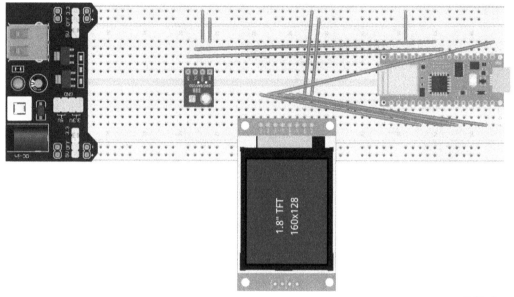

Figure 7.1 – Weather-station circuit (image is taken from Fritzing)

This is everything we need on assembling for the complete chapter. We can proceed to the next section and write the code that is able to *read the sensor data and display it* onto ST7735.

Programming the weather station

We are going to put the weather station logic, which consists of reading and interpreting sensor data, into a separate package so that we can use it in an example that only displays the data onto a display. When this is done, we are going to reuse this package to get the sensor data and calculate alerts to send it to an **MQTT broker**.

We start off by creating a new folder named Chapter07 inside the project folder, and create a new folder named weather-station inside Chapter07. We then create a new file named weather.go and name the package weatherstation. The project structure should now look like this:

Figure 7.2 – Project structure for programming the weather station

To implement the logic, follow these steps:

1. Define colors that are later to be used when we draw something on the display, as follows:

```
var (
    white = color.RGBA{255, 255, 255, 255}
    black = color.RGBA{0, 0, 0, 255}
)
```

2. Next, define an interface and insert the following functions. We are going to explain each function in detail as soon as we implement it, in some later steps of this list:

```
type Service interface {
    CheckSensorConnectivity()
    ReadData() (temperature, pressure, humidity int32,
        err error)
    DisplayData(temperature, pressure, humidity int32)
    GetFormattedReadings(temperature, pressure,
        humidity int32) (temp, press, hum string)
    SavePressureReading(pressure float64)
    CheckAlert(alertThreshold float64, timeSpan int8)
        (bool, float64)
}
```

3. We then define struct that contains the sensor and the display, as well as some more fields that we are going to explain as soon as we use them. For the BME280 device, we are going to use a driver from the TinyGo drivers repository. You can import it using the following path: tinygo.org/x/drivers/bme280. The code is shown in the following snippet:

```
type service struct {
    sensor *bme280.Device
    display *st7735.Device
    readings [6]float64
    readingsIndex int8
```

```
      firstReadingSaved bool
}
```

4. We then add a new constructor function that sets the sensor and display and
   initializes all values, as follows:

```
func New(sensor *bme280.Device,
    display *st7735.Device) Service {
        return &service{
        sensor: sensor,
        display: display,
        readingsIndex: int8(0),
        readings: [6]float64{},
        firstReadingSaved: false,
      }
    }
```

5. Then, add the ReadData function, which is a convenience function that reads all
   sensor values and returns them. The code can be seen in the following snippet:

```
func (service *service) ReadData() (temp, press, hum
    int32, err error) {temp, err =
        service.sensor.ReadTemperature()
            if err != nil {
                return
}
press, err = service.sensor.ReadPressure()
            if err != nil {
                return
}
hum, err = service.sensor.ReadHumidity()
            if err != nil {
                return
}
return
}
```

6. We then add a function that blocks the execution of the program until connection to the BME280 sensor has been approved, as follows:

```
func (service *service) CheckSensorConnectivity() {
    for {
        connected := service.sensor.Connected()
        if !connected {
            println("could not detect BME280")
            time.Sleep(time.Second)
    }
        println("BME280 detected")
        break
    }
}
```

7. We now add a function that takes the sensor readings and displays them on the screen, as follows:

```
func (service *service) DisplayData(
    temperature, pressure, humidity int32) {
```

8. Fill the screen so that we have no artifacts from previous calls. If we were to skip this step, we would possibly draw images above it that had previously been drawn, which would look really messy. The code is shown in the following snippet:

```
service.display.FillScreen(black)
```

9. Write a header using tinyfont, as follows:

```
tinyfont.WriteLineRotated(service.display,
    &freemono.Bold9pt7b, 110, 3, "Tiny Weather",
        white, tinyfont.ROTATION_90)
```

10. Convert the readings to strings, as follows:

```
temp, press, hum :=
    service.GetFormattedReadings(temperature,
        pressure, humidity)
```

11. Build and display the temperature, pressure, and humidity strings, as follows:

```
tempString := "Temp:" + temp + "C"
tinyfont.WriteLineRotated(service.display,
    &freemono.Bold9pt7b, 65, 3, tempString,
    white,tinyfont.ROTATION_90)
pressString := "P:" + press + "hPa"
tinyfont.WriteLineRotated(service.display,
    &freemono.Bold9pt7b, 45, 3, pressString, white,
    tinyfont.ROTATION_90)
humString := "Hum:" + hum + "%"
tinyfont.WriteLineRotated(service.display,
    &freemono.Bold9pt7b, 25, 3, humString, white,
    tinyfont.ROTATION_90)
}
```

12. Add a function that converts the sensor readings to °C, **hectopascal (hPa)**, and relative humidity as percentage to the strings, as follows:

```
func (service *service)
    GetFormattedReadings(
    temperature, pressure, humidity int32) (temp,
        press, hum string) {
temp = strconv.FormatFloat(
    float64(temperature/1000), 'f', 2, 64)
press = strconv.FormatFloat(
    float64(pressure/100000), 'f', 2, 64)
hum = strconv.FormatFloat(
    float64(humidity/100), 'f', 2, 64)
return
}
```

We have now finished implementing the logic to read and display the sensor data. The next step is calculating the weather alerts.

Calculating weather alerts

In order to calculate alerts, we need to save some readings. We can do so by following these steps:

1. For the weather-alert calculation, we will only need pressure. That is why we hold an array of float64 in the service struct, as can be seen in the following code snippet:

```
func (service *service) SavePressureReading(
    pressure float64) {
```

2. If we have ever saved a value before, we fill the complete array with the same value. This prevents some edge cases later on when calculating alerts. The code is shown in the following snippet:

```
if !service.firstReadingSaved {
    for i := 0; i < len(service.readings); i++ {
        service.readings[i] = pressure
}
```

3. As we have now inserted our first reading, we can set the true flag and return. This ensures that we only execute the preceding logic once. The code is shown in the following snippet:

```
service.firstReadingSaved = true
service.readingsIndex = 0
return
}
```

4. Store the reading into the current index. If our current index exceeds the maximum number of stored datasets, we reset the index; so, the next reading is going to overwrite the reading in index 0. The code is shown in the following snippet:

```
service.readingsIndex++
service.readingsIndex = service.readingsIndex %
    int8(len(service.readings))
service.readings[service.readingsIndex] = pressure
}
```

5. Add a function that uses the saved readings, calculates a difference between the two of them, and alerts if the difference exceeds the threshold. We are going to talk about thresholds and timespans later in this section when we call this function. The code is shown in the following snippet:

```
func (service *service) CheckAlert(alertThreshold
    float64, timeSpan int8) (bool, float64) {
        currentReading :=
            service.readings[service.readingsIndex]
```

6. Calculate the `comparisonIndex` value based on the `timeSpan` value, as follows:

```
currentReading := service.readings[currentIndex]
comparisonIndex := currentIndex - timeSpan
if comparisonIndex < 0 {
    comparisonIndex = 5 + comparisonIndex
}
```

7. Calculate the difference between the two values and raise an alert if the difference is greater than the threshold by returning `diff`, as follows:

```
comparisonReading := service.readings[comparisonIndex]
diff := comparisonReading - currentReading
return diff >= alertThreshold, diff
}
```

Okay—we just implemented an **application programming interface (API)** that lets us read, convert, and display sensor data, and along with it we can save sensor readings and calculate weather alerts.

Now, let's try out whether the code is actually able to read and display sensor data. To do so, we first create a new folder named `weather-station-example` inside the `Chapter07` folder. We then create a new `main.go` file with an empty `main` function inside. The project structure should now look like this:

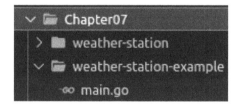

Figure 7.3 – Project structure for reading the code and displaying sensor data

Now, follow these steps to implement the example:

1. Inside the `main` function, we sleep for 5 seconds to get enough time to open PuTTY so that we are able to monitor the output on the serial port. The code is shown in the following snippet:

    ```
    time.Sleep(5 * time.Second)
    ```

2. Initialize and configure the display, as follows:

    ```
    machine.SPI0.Configure(machine.SPIConfig{
        Frequency: 12000000,
    })
    resetPin := machine.D6
    dcPin := machine.D5
    csPin := machine.D7
    backLightPin := machine.D2

    display := st7735.New(
        machine.SPI0, resetPin, dcPin, csPin,
            backLightPin)
    display.Configure(st7735.Config{
        Rotation: st7735.ROTATION_180,
    })
    ```

3. Initialize and configure the sensor. The sensor needs to be calibrated, which is done inside the `Configure` function. The code is shown in the following snippet:

    ```
    machine.I2C0.Configure(machine.I2CConfig{})
    sensor := bme280.New(machine.I2C0)
    sensor.Configure()
    ```

4. Create a new instance of `weatherstation` and wait for sensor connectivity. The code is shown in the following snippet:

    ```
    weatherStation := weatherstation.New(
        &sensor, &display)
    weatherStation.CheckSensorConnectivity()
    ```

5. Read and display the data, as follows:

```
for {
temperature, pressure, humidity, err :=
    weatherStation.ReadData()
if err != nil {
    println("could not read sensor data:", err)
    time.Sleep(1 * time.Second)
    continue
}
weatherStation.DisplayData(
    temperature, pressure, humidity)
time.Sleep(2 * time.Second)
}
```

That's it for this example. Now, go on and flash the program, using the following command:

```
tinygo flash --target=arduino-nano33 ch7/weather-station-
example/main.go
```

After a brief moment, the display should now look similar to the one shown here:

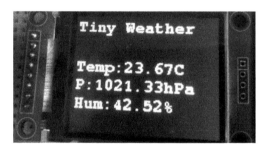

Figure 7.4 – Display output

We have now verified that we are able to read and display the sensor data. As we have now learned how to use the BMP280 sensor and have prepared a package that is able to calculate weather alerts, we can now go on to the next section, where we learn how to communicate with the Wi-Fi chip and how to send MQTT messages.

Sending MQTT messages to a broker

Let's now start to dive into the world of IoT. As every device that has a connection to the internet—or at least to some network—can be considered an IoT device, the project in this section can be considered an IoT project. The Arduino Nano 33 IoT has a `u-blox NINA-W102` chip on board that is capable of Wi-Fi communication. We can communicate with this chip using the SPI interface. As a driver for the NINA chip already exists, we don't have to implement one ourselves.

So, our plan is to send data through SPI to the NINA chip, which then sends the data through the network to an MQTT broker. The following diagram illustrates the process:

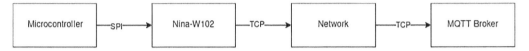

Figure 7.5 – Communication diagram

Although the driver functionality is wrapped in a package, some boilerplate code is still needed to start using the Wi-Fi chip. So, let's wrap it inside a new package.

Implementing the Wi-Fi package

We are going to create an API that provides functionality to initialize the NINA chip, check the hardware and set up a connection. So, let's start by creating a new folder named `wifi` inside the `Chapter07` folder and creating a new `wifi.go` file inside the newly created folder, and name the package `wifi`. The project structure should now look like this:

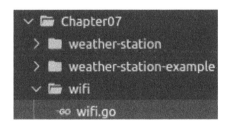

Figure 7.6 – Project structure

Now, perform these steps to implement the logic:

1. Define an interface for the package, as follows:

```go
type Client interface {
    Configure() error
    CheckHardware()
    ConnectWifi()
}
```

2. Add a client that stores the credentials as well as the SPI bus and `wifinina.Device`, as follows:

```go
type client struct {
    ssid string
    password string
    spi machine.SPI
    wifi *wifinina.Device
}
```

3. Add a constructor function that sets the SPI bus and the credentials, as follows:

```go
func New(ssid, password string) Client {
    return &client{
        spi: machine.NINA_SPI,
        ssid: ssid,
        password: password,
    }
}
```

4. Add the `Configure` function, as follows:

```go
func (client *client) Configure() error {
```

5. Configure the NINA SPI bus using the default pins, as follows:

```go
err := client.spi.Configure(machine.SPIConfig{
    Frequency: 8 * 1e6,
    SDO: machine.NINA_SDO,
    SDI: machine.NINA_SDI,
    SCK: machine.NINA_SCK,
```

```
})
    if err != nil {return err
}
```

6. Create a new instance of the wifinina driver and pass the SPI bus as well as the default pins, as follows:

```
client.wifi = &wifinina.Device{
    SPI: client.spi,
    CS: machine.NINA_CS,
    ACK: machine.NINA_ACK,
    GPIO0: machine.NINA_GPIO0,
    RESET: machine.NINA_RESETN,
}
client.wifi.Configure()
```

7. The chip needs a moment before it is ready to be used, which is why we sleep for a brief moment. The code is shown in the following snippet:

```
time.Sleep(5 * time.Second)
return nil
}
```

8. Now, we add a function to check the hardware, as follows:

```
func (client *client) CheckHardware() {
```

9. First, we print the currently installed **firmware version**. This information can be important if you face any issues using the NINA chip. Also, you can use this information to check which features are supported by the firmware. The code is shown in the following snippet:

```
firmwareVersion, err := client.wifi.GetFwVersion()
if err != nil {
    return err
}
println("firmware version: ", firmwareVersion)
```

10. Now, we scan for available Wi-Fi networks and print all results. The internal buffer only stores up to 10 **service set identifiers** (**SSIDs**). If the scan for Wi-Fi networks runs without any errors, we can be sure that we are able to communicate with the chip. The code is shown in the following snippet:

```
result, err := client.wifi.ScanNetworks()
if err != nil {
    return err
}
for i := 0; i < int(result); i++ {
    ssid := client.wifi.GetNetworkSSID(i)
    println("ssid:", ssid, "id:", i)
}
}
```

11. Now, we implement a convenience function that establishes a connection to a network, as follows:

```
func (client *client) ConnectWifi() {
    println("trying to connect to network: ",
        client.ssid)
    client.connect()
    for {
```

12. Sleep for a second, as it can take a while until the connection is established. The code is shown in the following snippet:

```
time.Sleep(1 * time.Second)
```

13. Get the connection status and print it, as follows:

```
status, err := client.wifi.GetConnectionStatus()
if err != nil {
    println("error:",err.Error())
}
println("status:",status.String())
```

14. If the status equals `StatusConnected`, as shown in the following code snippet, we are successfully connected to the network:

```
if status == wifinina.StatusConnected {
    break
}
```

15. Sometimes, the connection cannot be established on the first attempt, which is why we just try it again, as illustrated in the following code snippet:

```
if status == wifinina.StatusConnectFailed ||
    status == wifinina.StatusDisconnected {
        client.connect()
    }
}
```

16. After the connection has successfully been established, we print the **Internet Protocol (IP)** address that our device has been assigned by the **Dynamic Host Configuration Protocol (DHCP)**, as follows:

```
ip, _, _, err := client.wifi.GetIP()
if err != nil {
    println("could not get ip address:", err.Error())
}
println("connected to wifi. got ip:", ip.String())
}
```

17. We can either only set the network (`ssid`) with no passphrase for open networks or we can set the network (`ssid`) and passphrase. Setting either of these options triggers a connection attempt. If no password has been set, we try to connect to an open network. If the password and `ssid` are set, we try to connect to a secured network, as follows:

```
func (client *client) connect() error {
    if client.password == "" {
        return client.wifi.SetNetwork(client.ssid)
    }
    return client.wifi.SetPassphrase(
```

```
        client.ssid, client.password)
    }
```

That's everything we need to implement our abstraction layer. We are going to test this package together with an MQTT client abstraction layer, which we are going to implement next.

Implementing an MQTT client abstraction layer

Just as with the Wi-Fi driver, the MQTT client needs some boilerplate code in order to get up and running. We are going to reduce the boilerplate code by adding an abstraction layer. That way, we only have the boilerplate code one time in a reusable component and do not have to write the same code repeatedly in every future project.

We start by creating a new folder named mqtt-client inside the Chapter07 folder, and create a new file named client.go and place it inside the mqttclient package. The project structure should now look like this:

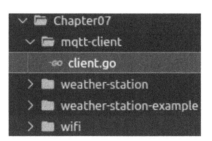

Figure 7.7 – Project structure

Before we start to implement the code, we first need to understand what MQTT is and how it works.

Understanding MQTT

MQTT is a messaging protocol for the IoT. It is based on a **publisher/subscriber** architecture. A microcontroller that reads sensor data can **publish messages** to a so-called **topic** (such a microcontroller would be a thing in the IoT world). These messages are sent to a **broker**.

The MQTT standard allows the usage of **Transport Layer Security** (**TLS**), as well as **Open Authorization** (**OAuth**) for authentication. It is also possible to not authenticate at all. The available authentication flows depend on the implementation and configuration of the used MQTT broker. Securing the broker by using authentication flows is important when sending sensitive data over the internet. The following diagram shows an example architecture of a single MQTT broker and multiple MQTT clients:

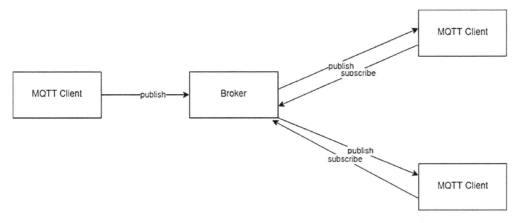

Figure 7.8 – MQTT architecture

In order to use MQTT, we need an active broker to which a client can publish messages. We also need one or many clients to be able to subscribe to messages from a certain topic.

The sequence diagram of a typical MQTT communication is straightforward and is based on a **command and command acknowledge** pattern. Let's have a look at an example sequence, as follows:

1. The client connects to the broker.

2. The broker acknowledges the connection.

3. Optional: The client subscribes to a topic.

4. The broker acknowledges the subscription.

5. Optional: The client publishes a message.

6. The broker acknowledges the published message.

This looks like the sequence shown in the following diagram:

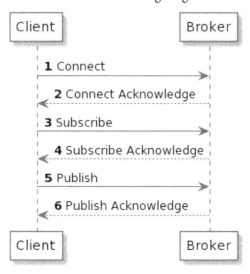

Figure 7.9 – MQTT sequence diagram (image was created using PlantUML)

Summarizing this, here we mean a broker can serve many clients, a client can subscribe to one or many topics, and a client can publish a message in a topic. This should be enough base knowledge about MQTT.

> **Note**
>
> If you want to gain a deeper knowledge about MQTT, you might want to have a look at the specification:
>
> `https://mqtt.org/mqtt-specification/`

Let's now write the abstraction layer. As we have already prepared the project structure, we can start directly writing the code by following these steps inside the `client.go` file:

1. As our client is only going to publish messages, our API is going to be fairly simple. We need to be able to connect to a broker and we need to be able to publish messages. We now add `struct` that contains the **Uniform Resource Locator (URL)** of the broker, an identifier that identifies us, and an `mqtt.Client` from the `drivers` repository, as follows:

```
type Client struct {
    mqttBroker string
    mqttClientID string
    MqttClient mqtt.Client
}
```

2. To create a new `Client`, we only need to set the `mqttBroker` URL, as follows:

```
func New(mqttBroker, clientID string) *Client {
    return &client{
        mqttBroker: mqttBroker,
        MqttClientID: clientID,
    }
}
```

3. Now, add the `ConnectBroker` function, which is going to establish a connection to the broker. The code is shown in the following snippet:

```
func (client *client) ConnectBroker() error {
```

4. We create new client options that will later be passed as arguments when creating a new client. These options take all the parameters needed to establish a connection. When using a broker that requires a password and username, these can also be set here. The code is shown in the following snippet:

```
opts := mqtt.NewClientOptions().
        AddBroker(client.mqttBroker).
```

When testing programs using a local broker, we sometimes try to connect while the old client connection has not been discarded and we could run into problems connecting with the same `clientID` again, so using random strings helps a lot. The MQTT specification states that a `clientID` should be between 1 and 23 characters long, but brokers such as Mosquitto do not implement that. We will learn about the Mosquitto MQTT broker later in this section.

ClientIDs must be unique—otherwise, clients would be kicked out by the broker.

5. We are going to use a combination of the `clientID` we passed in and a random string of length 4, as illustrated in the following code snippet:

```
SetClientID(client.mqttClientID + randomString(4))
```

6. We now create a new client and pass `opts` as a parameter, and try to connect to the broker as follows:

```
client.mqttClient = mqtt.NewClient(opts)
token := client.MqttClient.Connect()
```

7. Although the current implementation of the token always return `true` when using the `wait` function, we still add it here in case it is implemented by the time you have worked through this chapter. We can use this function to wait for any command to get *acked* (acknowledged). Alternatively, we could use `token.WaitTimeout`, which internally times out when the given time span is over. The first option is shown in the following code snippet:

```
if token.Wait() && token.Error() != nil {
    return token.Error()
}
return nil
}
```

8. Add the `PublishMessage` function. qos (the **quality of service (QOS)**) can be 0, 1, or 2. After we have completely implemented this package, we are going to have a deeper look at the qos levels. The `retain` flag tells the broker to store the last message having a `retain` flag. When a new client subscribes to the broker, the retained message will directly get delivered. The code is shown in the following snippet:

```
func (client *client) PublishMessage(
    topic string, message []byte, qos uint8, retain
        bool) error {
    token := client.MqttClient.Publish(
        topic, qos, retain, message)
    if token.WaitTimeout(time.Second) &&
        token.Error() != nil {
            return token.Error()
    }
    return nil
}
```

9. The next step is to add a function that allows us to subscribe to a certain topic. The following code snippet illustrates this:

```
func (client *Client) Subscribe(
    topic string, qos byte, callback mqtt.MessageHandler)
        error {
    token :=
    client.MqttClient.Subscribe( topic, qos, callback)
```

```
        if token.WaitTimeout(time.Second)
            && token.Error() != nil {
                return token.Error()
    }
    return nil
    }
```

10. Now, add a function to generate a random string containing a character between A and Z. The following functions are taken from the mqtt driver example:

```
func randomInt(min, max int) int {
    return min + rand.Intn(max-min)
}
func randomString(len int) string {
    bytes := make([]byte, len)
    for i := 0; i < len; i++ {
        bytes[i] = byte(randomInt(65, 90))
    }
    return string(bytes)
}
```

That's it for our abstraction layer. Before we go on to write the actual weather-station program, let's have a look at the QOS levels.

MQTT provides three QOS levels, which work as follows:

- **QOS 0**: *A message is delivered once.* The message is not stored by the sender and is not acknowledged. So, the client will only try to deliver it once, and if that fails, the message is discarded.

- **QOS 1**: *The message is delivered at least once.* It is guaranteed that the message is being sent to the broker. The client tries to resend the message with an additional duplicate flag until it gets acknowledged by the broker. All of these messages will be sent to subscribed clients.

- **QOS 2**: *The message will be delivered only once.* As in **QOS 1**, the message will be resent with a duplicate flag until the message has been acknowledged. The difference is that the message will only be delivered to subscribers when the client sends a PUBREL (publish release) message.

You can find more information about the underlying processes by following this link:

`https://www.hivemq.com/blog/mqtt-essentials-part-6-mqtt-quality-of-service-levels/`

As we now have a basic understanding of MQTT and have implemented our abstraction layer, it's time to put everything together in the next step and actually start to publish messages to a broker.

Implementing the weather station

We have prepared all the code that we need to implement the actual logic but we do not have an MQTT broker yet. So, let's set up a local MQTT broker that we can use for this program.

Setting up an Eclipse Mosquitto MQTT broker

We are going to use the Eclipse Mosquitto MQTT broker. You can find more information regarding the broker here: `https://mosquitto.org/`.

If you do not want to set up a local MQTT broker or if you cannot use Docker right now, you can skip this and use the Mosquitto test system. *But please only use the Mosquitto test system for testing purposes; also, never publish any sensitive data to the test system as anyone could listen to the messages.* You can find the needed URLs and ports for the test system here: `http://test.mosquitto.org/`.

It is also possible to install a Mosquitto broker locally without using Docker, but that process won't be covered in this book as using Docker is an easy and straightforward process, while setting up Mosquitto locally is more complicated. To set up Mosquitto using Docker, we need to create a configuration file. To do so, create a new folder named `mosquitto` inside the `Chapter07` folder and create a new folder named `config` inside the `mosquitto` folder. Now, create a new file and name it `mosquitto.conf`. The next step is to insert the following configuration:

```
user mosquitto
listener 9001 127.0.0.1
protocol websockets
allow_anonymous true
```

We have configured Mosquitto to use the user `mosquitto` and listen on all IP addresses of the host. We also listen for Websocket connections on the localhost using port `9001`, which we make use of in the Wasm app in the *Implementing the weather app* section later in this chapter.

The `allow_anonymous` flag allows unauthenticated clients to connect. For more information regarding the possible configuration options, consult the main manual page at `https://mosquitto.org/man/mosquitto-conf-5.html`.

Now, we only need to start the container. It is important to map ports `1883` and `9001` so that we can actually reach these ports. Also, we need to *pass the path to our config file. So, replace the path that I used with the actual path to the config file on your system*, as follows:

```
docker run -it --name mosquitto \
--restart=always \
-p 1883:1883 \
-p 9001:9001 \
-v ~/go/src/ github.com/PacktPublishing/Creative-DIY-
Microcontroller-Projects-with-TinyGo-and-WebAssembly /
Chapter07/mosquitto/config/mosquitto.conf:/mosquitto/config/
mosquitto.conf:ro \
eclipse-mosquitto
```

As we now have a running Mosquitto instance, we can now truly begin to implement our client logic.

Let's start by creating a new folder named `weather-station-mqtt` inside the `Chapter07` folder, then create a new `main.go` file with an empty `main` function inside. The project structure should now look like this:

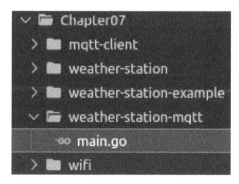

Figure 7.10 – Project structure for weather-station-mqtt

When developing these examples, I have faced some issues regarding the `wifinina` driver inside the TinyGo `drivers` repository. I am currently working on resolving these issues. So, if you face any problems connecting to your Wi-Fi network, use the two imports shown in the following code snippet instead of the official ones. You will also need to change these imports in the `wifi` and `mqtt-client` packages we developed earlier in this chapter:

```
github.com/Nerzal/drivers/wifinina
github.com/Nerzal/drivers/net/mqtt
```

Also, when using my fork of the `drivers` repository for the `wifinina` driver, the initialization of `wifi.Client` looks a little bit different. When using the driver, you will see that there is an error in the `Configure` function. To fix it, replace the initialization of the `client.wifi` object with the following snippet:

```
wifiDevice := wifinina.NewSPI(
    client.spi,
    machine.NINA_CS,
    machine.NINA_ACK,
    machine.NINA_GPIO0,
    machine.NINA_RESETN,
)
client.wifi = wifiDevice
```

To implement the program, follow these steps:

1. Define constants for `ssid` and `password`, as shown in the following code snippet. You must insert your own credentials here:

```
const ssid = "changeMe"
const password = "changeMe"
```

2. Define variables for `temperature`, `pressure`, and `humidity`, as shown in the following code snippet. These will be accessed by multiple goroutines:

```
var (
    temperature float64
    pressure float64
    humidity float64
)
```

3. When watching streams from Ron Evans (who is one of the TinyGo maintainers), I learned a helpful trick. If an error occurs when doing really important things, we want to be able to find the error message in the serial output. In such a case, we pause the program execution and repeatedly print the message, as follows:

```
func printError(message string, err error) {
    for {
        println(message, err.Error())
        time.Sleep(time.Second)
    }
}
```

4. Now, inside the `main` function, we start by sleeping a brief moment so that we have enough time to open up PuTTY to monitor the serial output while we initialize the sensor and weather station, as follows:

```
time.Sleep(5 * time.Second)
```

5. Initialize `weatherStation`, as follows:

```
machine.I2C0.Configure(machine.I2CConfig{})
sensor := bme280.New(machine.I2C0)
sensor.Configure()
weatherStation := weatherstation.New(&sensor, nil)
weatherStation.CheckSensorConnectivity()
```

6. Create a new `wifi` client, as follows:

```
wifiClient := wifi.New(ssid, password)
```

7. Configure the `wifi` client, as follows:

```
println("configuring nina wifi chip")
err := wifiClient.Configure()
if err != nil {
    printError("could not configure wifi client", err)
}
```

8. Now, we call the `CheckHardware` function, which will output the firmware version and `ssids` that have been found when scanning for networks. If that works, we can be sure that the microcontroller is able to communicate with the NINA chip. The code is shown in the following snippet:

```
println("checking firmware")
err = wifiClient.CheckHardware()
if err != nil {
    printError("could not check hardware", err)
}
```

9. Try to connect to the network, as follows:

```
wifiClient.ConnectWifi()
```

Create a new `mqttClient` instance. Please note that you have to change the IP address to the address of the host where your MQTT broker is running. *Do not omit the* `tcp://` *part*, as it is being used by the driver implementation to estimate which kind of connection needs to be established. The code is shown in the following snippet:

```
mqttClient := mqttclient.New("tcp://192.0.2.22:1883")
```

10. Try to connect to the MQTT broker, as follows:

```
println("connecting to mqtt broker")
err = mqttClient.ConnectBroker()
if err != nil {
    printError("could not configure mqtt", err)
}
println("connected to mqtt broker")
```

11. Start a goroutine that publishes sensor data, as follows:

```
go publishSensorData(
    mqttClient, wifiClient, weatherStation)
```

12. Start a goroutine that publishes weather alerts, as follows:

```
go publishAlert(
    mqttClient, wifiClient, weatherStation)
```

13. Read new sensor values once a minute, as follows:

```
for {
    temperature, pressure, humidity, err
        = weatherStation.ReadData()
        if err != nil {
        printError("could not read sensor data:", err)
}
time.Sleep(time.Minute)
}
```

14. Now, add the `publishSensorData` function. For testing purposes it runs once per minute, but you can customize it depending on your needs. The code is shown in the following snippet:

```
func publishSensorData(mqttClient mqttclient.Client,
    wifiClient wifi.Client, weatherStation
    weatherstation.Service) {
    for {
        time.Sleep(time.Minute)
        println("publishing sensor data")
        tempString, pressString,
        humidityString:=weatherStation.
        GetFormattedReadings(temperature, pressure,
        humidity)
```

15. As most encoding packages are currently not supported by TinyGo, we use a **character-separated string** to serialize the data, as this will be easy to deserialize on the subscriber side. The code is shown in the following snippet:

```
message := []byte(fmt.Sprintf("sensor
    readings#%s#%s#%s", tempString, pressString,
        humidityString))
```

Sometimes, we lose the connection to Wi-Fi or to the MQTT broker. In that case, we just try to establish a new connection, as follows:

```
err := mqttClient.PublishMessage("weather/data",
message, 0, true)
if err != nil {
    switch err.(type) {
```

```
        println(err.Error())
    case wifinina.Error:
        wifiClient.ConnectWifi()
        mqttClient.ConnectBroker()
    default:
        println(err.Error())
    }
  }
 }
}
```

16. We now add the `publishAlert` function, which runs once an hour, as follows:

```
func publishAlert(mqttClient mqttclient.Client,
    wifiClient wifi.Client, weatherStation
    weatherstation.Service) {
        for {
            time.Sleep(time.Hour)
```

17. We save the pressure reading on an hourly basis, as follows:

```
weatherStation.SavePressureReading(pressure)
```

18. Now, we check for whether we have to send an alert. We use 2 as the value for the alert threshold, for the hourly check. We will talk about these values in more detail after finishing the implementation of the function. The code is shown in the following snippet:

```
alert, diff := weatherStation.CheckAlert(2, 1)
```

19. If we have an `alert`, we publish it by running the following code:

```
if alert {
    err := mqttClient.PublishMessage("weather/alert",
        []byte(fmt.Sprintf("%s#%v#%s", "possible storm
        incoming", diff, "1 hour")), 0, true)
    if err != nil {
        switch err.(type) {
        case wifinina.Error:
            println(err.Error())
            wifiClient.ConnectWifi()
```

```
            mqttClient.ConnectBroker()
        default:
            println(err.Error())
        }
    }
}
```

20. Now, we check for an `alert` on a 3-hour schedule, as follows:

```
alert, diff = weatherStation.CheckAlert(6, 3)
```

If we do not have an `alert`, we continue by running the following code:

```
if !alert {
    continue
}
```

21. Publish the alert (if we have one) on a 3-hour schedule, as follows:

```
err := mqttClient.PublishMessage("weather/alert",
    []byte(fmt.Sprintf("%s#%v#%s", "possible storm
    incoming", diff, "3 hours")), 0, true)
if err != nil {
    println(err.Error())
    switch err.(type) {
        case wifinina.Error:
            wifiClient.ConnectWifi()
            mqttClient.ConnectBroker()
        }
    }
}
```

That's everything we need for our first IoT project. We have now developed a client that reads data from a sensor and publishes it to an MQTT broker. The client also checks the data for possible incoming storms and publishes these warnings as messages on a different topic. Before we try out the program, let's briefly talk about the thresholds and timespans we used as parameters in the alerts.

> **Very important note**
>
> I am by no means a meteorologist. *This program is not able to predict every possible incoming storm.* The data I used as example values is not tested as I experienced no incoming storms when writing this book. Also, this information is based on articles I read online and might only work quite well in the place I'm living, so you might need to do your own research about the coherence of pressure changes and incoming bad weather. If this program did not predict an incoming storm but you feel one could be incoming, please check your local news and weather sources for that information. *So, again, if you live in an area that is frequently being hit by dangerous storms, please do not blindly trust this program.* Meteorology is way more complicated than this program and we do only check for incoming storms based on sudden drops in pressure. There are more indicators for incoming storms, so please consider this program a prototype.

The source of my threshold values is the following web page:

`http://www.bohlken.net/airpressure2.htm`

This states that a pressure drop of >=**2hPa** during a 1-hour period could indicate a possible incoming storm. It also states that a pressure drop of >=**6hPa** during a 3-hour period could indicate a possible incoming storm.

If you are interested in MQTT best practices, check out the following link:

`https://www.hivemq.com/blog/mqtt-essentials-part-5-mqtt-topics-best-practices/`

We can now go on and finally flash the program onto the microcontroller by using the following command:

```
tinygo flash --target=arduino-nano33 ch7/weather-station-mqtt/
main.go
```

We have now flashed the program and everything seems to be running fine, but how do we know that the messages are really being successfully published? We can use an MQTT **graphical user interface** (**GUI**) client in such cases. I do recommend **MQTT Explorer** for this. You can download the program for every platform here:

`https://mqtt-explorer.com/`

After starting the program, you only need to insert the `hostname` and `port` values. Just use `localhost` and port `1883` as parameters, then save your connection. The connection window should now look similar to this:

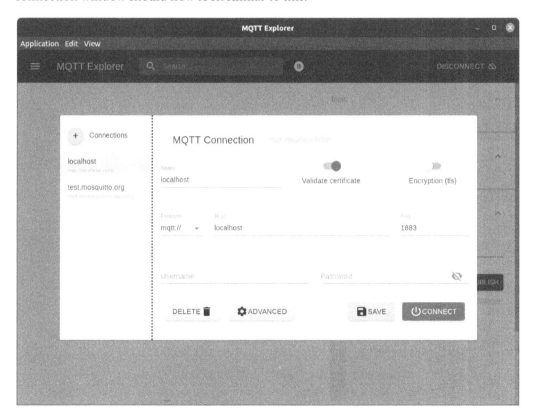

Figure 7.11 – MQTT Explorer connection window

When the program is running on the microcontroller, you should be able to see the topics and messages that are being published to the broker. This will look similar to the following output:

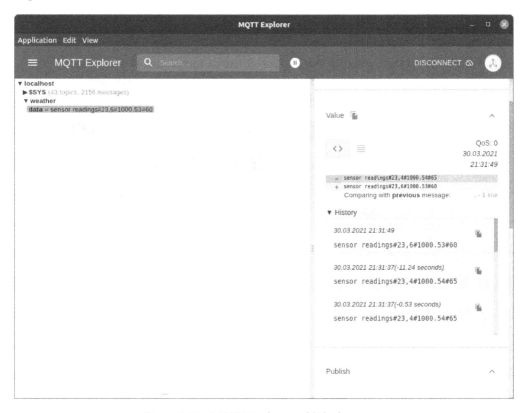

Figure 7.12 – MQTT Explorer published message

We have learned how to implement a weather station that is able to publish messages to different topics to an MQTT broker. We have also learned how to set up Mosquitto using Docker, and we have learned how to use MQTT Explorer to check if our messages really are being published. The next step is to create a Wasm app that displays these messages.

Introducing Wasm

Let's find out what Wasm is. The *WebAssembly* home page states the following:

> *"WebAssembly (abbreviated Wasm) is a binary instruction format for a stack-based virtual machine. Wasm is designed as a portable compilation target for programming languages, enabling deployment on the web for client and server applications."*

Source: `https://webassembly.org/`

In other words, we can write code in any language and compile it to the Wasm binary format, which can then be executed by the browser. That makes Wasm extremely valuable, as we can create client applications using languages other than JavaScript.

A great advantage of Wasm compared to JavaScript is that it aims to execute at native speed, and as it runs in a sandboxed environment inside the browser it can be considered as relatively safe. Luckily, TinyGo does support Wasm as a compilation target, so we can make use of the incredibly small binary sizes that TinyGo produces, which will significantly speed up page loading times compared to other technologies.

Here is a list to summarize the preceding information:

- **Wasm is a new language** that officially became the fourth language of the web, after **HyperText Markup Language** (**HTML**), **Cascading Style Sheets** (**CSS**), and JavaScript. Source: `https://www.w3.org/2019/12/pressrelease-wasm-rec.html.en`
- **Wasm is a binary format**. It does not aim to be human-readable.
- **Wasm is so low-level** that it brings performance improvements compared to high-level languages such as JavaScript.
- Go code can be compiled in the Wasm binary format.

As we now have a brief basic understanding of what Wasm theoretically is, let's just use it to get a better understanding.

Displaying sensor data and weather alerts on a Wasm page

Our goal is to develop a small application that displays weather alerts and our sensor data that is being published to an MQTT broker, so we will need some very basic HTML and JavaScript skills in order to achieve this. We start by developing a small server that serves the Wasm app to a client.

Serving the application

As Wasm is served to a browser, we need an HTTP endpoint that serves all files we might need. We start by creating a new folder named wasm-server inside the Chapter07 folder, and inside this folder we create a new main.go file with an empty main function. Now, follow these steps to implement the server:

1. Define the directory where the FileServer should look for files, as follows:

    ```
    const dir = "Chapter07/html"
    ```

2. Now, inside the main function, create a new FileServer and pass the directory as a parameter, as follows:

    ```
    fs := http.FileServer(http.Dir(dir))
    ```

3. Start an HTTP server that listens on port 8080, as follows:

    ```
    err := http.ListenAndServe(":8080",
            http.HandlerFunc(func(resp http.ResponseWriter,
            req *http.Request) {
    ```

4. Tell the browser not to cache the files. *We only use this for our development environment* to ensure that we never get a cached version of the app. On production, you would not want to deactivate the cache. The code is shown in the following snippet:

    ```
    resp.Header().Add("Cache-Control", "no-cache")
    ```

5. We need to set the proper content-type header for different file types, as follows:

    ```
    if strings.HasSuffix(req.URL.Path, ".wasm") {
        resp.Header(). Set("content-type",
            "application/wasm")
    }
    ```

6. Now, finally, just let the file server serve the files, as follows:

```
fs.ServeHTTP(resp, req)
}))
if err != nil {
    println("failed to server http requests:",
        err.Error())
}
```

This is everything we need to prepare our server. The server is now able to serve all files from inside the `html` directory, which we are going to create later. Note that this example server is nearly exactly the same as the example server from the TinyGo repository.

Implementing the weather app

Let's now create the actual app. The app is going to consist of three parts, as follows:

1. An HTML file, which is going to hold the page content.
2. A `wasm.js` file, which is going to execute the Go code and also hold some other helper functions.
3. A `wasm_exec.js` file, which can be considered as glue code, as it maps Go functions to JavaScript functions. This file is going to be provided by TinyGo itself.

Let's start to create our HTML file. To do so, create a new folder named `weather-app` inside the `Chapter07` folder and create a new file named `index.html` inside this new folder. Now, follow these steps inside the HTML file:

1. Declare metadata such as `charset`, `title`, and `viewport` in the head, as follows:

```
<!DOCTYPE html>
<html>
<head>
    <meta charset="utf-8" />
    <title>TinyGo Weather Station</title>
    <meta name="viewport" content="width=device-width,
    initial-scale=1" />
```

2. Import the `paho mqtt` library. We are going into more detail on this one as soon as we use it. The code is shown in the following snippet:

```
<script src="https://cdnjs.cloudflare.com/ajax/libs/paho-
mqtt/1.0.1/mqttws31.min.js" type="text/javascript"></
script>
```

3. Import the `wasm_exec.js` and `wasm.js` files. These files are being provided by our server. The code is shown in the following snippet:

```
<script src="wasm_exec.js"
    type="text/javascript"></script>
<script src="wasm.js" type="text/javascript"></script>
</head>
```

4. Now, in the body, we want to tell the user what the app is about and display our data, as follows:

```
<body>
    <h1>TinyGo Weather Station</h1>
    <p>Alerts:</p>
```

5. Define a table with four columns that are going to be dynamically filled with our weather alerts. The `tbody` column gets an `id` attribute so that we are able to identify that element. The code is shown in the following snippet:

```
<table>
    <thead>
        <tr>
            <th>TimeStamp</th>
            <th>Message</th>
            <th>Pressure Difference</th>
            <th>Time span</th>
        </tr>
    </thead>
    <tbody id="tbody-alerts"></tbody>
</table>
```

6. Define a table with five columns that are going to be filled with our sensor data. The tbody column again gets an id. The code is shown in the following snippet:

```
<p>Events:</p>
<table>
    <thead>
        <tr>
            <th>TimeStamp</th>
            <th>Message</th>
            <th>Temperature</th>
            <th>Pressure</th>
            <th>Humidity</th>
        </tr>
    </thead>
    <tbody id="tbody-data"></tbody>
</table>
</body>
</html>
```

This is everything we need to display our data. I did not include any CSS, to keep the example as easy to understand and short as possible. You can, of course, also include inline CSS or reference a CSS file that could also be served from the html directory.

Reading more

If you want to learn how to create beautiful web apps using HTML5 and CSS, I recommend the book *Responsive Web Design with HTML5 and CSS* by the awesome author Ben Frain. You can find it at the following link:

```
https://www.packtpub.com/product/responsive-
web-design-with-html5-and-css-third-
edition/9781839211560
```

The next step is to implement the actual client logic. We start by creating a new file inside the weather-app directory and naming it wasm.go, and also create an empty main function inside the newly created file. Now, follow these steps:

1. Define a struct for our sensor events, as follows:

```
type sensorEvent struct {
    TimeStamp string
```

```
        Message string
        Temperature float32
        Pressure float32
        Humidity float32
    }
```

2. Define a struct for our alert events, as follows:

```
    type alertEvent struct {
        TimeStamp string
        Message string
        Diff string
        TimeSpan string
    }
```

3. Create a channel that is going to be used to handle the sensor events, as follows:

```
    var sensorEvents = make(chan sensorEvent)
```

4. Create a channel that is going to be used to handle the alert events, as follows:

```
    var alertEvents = make(chan alertEvent)
```

5. Inside the `main` function, we export the `sensorDataHandler` function as `sensorDataHandler` to the JavaScript environment. This way, we can call the `go` function from JavaScript. The code is shown in the following snippet:

```
    js.Global().Set("sensorDataHandler",
        js.FuncOf(sensorDataHandler))
```

6. We also export the `alertHandler` function, as follows:

```
    js.Global().Set("alertHandler",
        js.FuncOf(alertHandler))
```

7. Start a goroutine that handles the sensor events, as follows:

```
    go handleSensorEvents()
```

8. Start a goroutine that handles the alert events, as follows:

```
    go handleAlertEvents()
```

9. Block the execution of the main goroutine so that the program does not just shut down after executing the main function, as follows:

```
wait := make(chan struct{}, 0)
<-wait
```

10. Add the alertHandler function. In order to be able to export the function using the js.Global().Set() call, the function must have a signature that accepts a js.Value and a [].js.Value and returns an interface{}, as illustrated in the following code snippet:

```
func alertHandler(this js.Value, args []js.Value)
    interface{} {
```

11. When calling this function, we pass a single string as parameter. We will be able to find the string inside the first index of args. Afterward, we need to split the message using a hashtag as separator. The code is shown in the following snippet:

```
message := args[0].String()
splittedStrings := strings.Split(message, "#")
```

12. Add the deserialized message to the channel. As we have not placed a timestamp onto the message when sending it, we now add the timestamp, as illustrated in the following code snippet:

```
alertEvents <- alertEvent{
    TimeStamp: time.Now().Format(time.RFC1123),
    Message: splittedStrings[0],
    Diff: splittedStrings[1],
    TimeSpan: splittedStrings[2],
}
```

13. Simply return nil as we do not need to write back any value to the JavaScript code that calls this function, as illustrated in the following code snippet:

```
return nil
}
```

14. We now do the same procedure for the sensor data events, as follows:

```
func sensorDataHandler(this js.Value, args []js.Value)
    interface{} {
message := args[0].String()
splittedStrings := strings.Split(message, "#")
sensorEvents <- sensorEvent{
    TimeStamp: time.Now().Format(time.RFC1123),
    Message: splittedStrings[0],
    Temperature: splittedStrings[1],
    Pressure: splittedStrings[2],
    Humidity: splittedStrings[3],
}
return nil
}
```

15. Add the `handleAlertEvents()` function. This function loops forever and reads an alert from the channel. The code is shown in the following snippet:

```
func handleAlertEvents() {
for {
    event := <-alertEvents
```

16. As we have read an alert event, we need to find the `tbody` element in the `html` directory in order to add a new row. We make use of some helper functions that we are going to explain as soon as we implement them. The code is shown in the following snippet:

```
tableBody := dom.GetElementByID("tbody-alerts")
```

17. Create a new table row, as follows:

```
tr := dom.CreateElement("tr")
```

18. Add the column data, as follows:

```
dom.AddTd(tr, event.TimeStamp)
dom.AddTd(tr, event.Message)
```

19. Add the formatted column data, as follows:

```
dom.AddTdf(tr, "%s hPa", event.Diff)
dom.AddTdf(tr, "%s", event.TimeSpan)
```

20. Append a new `tableRow` to `tbody`, as follows:

```
dom.AppendChild(tableBody, tr)
println("successfully added sensor event to
    table")
    }
}
```

21. The `handleSensorEvents` function works in a very similar way. We loop forever, read events from the `sensorEvents` channel, and add the data to the `tbody`. The code is shown in the following snippet:

```
func handleSensorEvents() {
for {
    event := <-sensorEvents
    tableBody := dom.GetElementByID("tbody-data")
    tr := dom.CreateElement("tr")
    dom.AddTd(tr, event.TimeStamp)
    dom.AddTd(tr, event.Message)
    dom.AddTdf(tr, "%s°C", event.Temperature)
    dom.AddTdf(tr, "%s hPa", event.Pressure)
    dom.AddTdf(tr, "%s", event.Humidity)
    dom.AppendChild(tableBody, tr)
    println("successfully added sensor event to
    table")
    }
}
```

The only thing missing from our Go code is the dom helper function. So, create a new folder named dom inside the Chapter07 folder, create a new dom.go file inside the folder, and name the package dom. Now, follow these steps to implement it:

1. Add a GetDocument function that wraps the get document call. You can also refer to the Getdocument as the HTML. The code is shown in the following snippet:

    ```
    func GetDocument() js.Value {
        return js.Global().Get("document")
    }
    ```

2. Add a wrapper for the createElement call. A created element is not directly visible. A newly created element needs to be added to the document before it is being rendered. The code is shown in the following snippet:

    ```
    func CreateElement(tag string) js.Value {
        document := GetDocument()
        return document.Call("createElement", tag)
    }
    ```

3. Add a wrapper for the getElementById function. We used this function to get the tbody elements, using the id we defined in the html directory. The code is shown in the following snippet:

    ```
    func GetElementByID(id string) js.Value {
        document := GetDocument()
        return document.Call("getElementById", id)
    }
    ```

4. Add a wrapper for appendChild. We used this function to add the tableRows into the tbody elements. This actually adds the elements to the html directory. The code is shown in the following snippet:

    ```
    func AppendChild(parent js.Value, child js.Value) {
        parent.Call("appendChild", child)
    }
    ```

5. Add a wrapper to set the `innerHTML` function. This function adds the given value between `html` tags. The code is shown in the following snippet:

```
func SetInnerHTML(object js.Value, value interface{}) {
    object.Set("innerHTML", value)
}
```

6. The `AddTd` function creates a new `td` element, sets the `innerHTML` function, and appends the child to the given `tr` element, as illustrated in the following code snippet:

```
func AddTd(tr js.Value, value interface{}) {
    td := CreateElement("td")
    SetInnerHTML(td, value)
    AppendChild(tr, td)
}
```

7. The `AddTdf` function does the same as the `AddTd` function, with the difference that the `innerHTML` function gets formatted. The code is shown in the following snippet:

```
func AddTdf(tr js.Value, formatString string, value
    interface{}) {
    td := CreateElement("td")
    SetInnerHTML(td, fmt.Sprintf(formatString, value))
    AppendChild(tr, td)
}
```

We have now implemented all the helper functions that we used in the wasm.go file. The only thing missing, before we can build and test the app, is the wasm.js file. So, let's create a new file named wasm.js and follow these steps to implement the last part:

1. Declare that this file should be executed in strict mode. For more information on strict mode, check out the following site:

   https://www.w3schools.com/js/js_strict.asp

2. Define a `const` value for the `wasm` file. The binary we build will later be named `wasm.wasm`. We also add new variables to store the `mqtt` client and the `wasm` object, as illustrated in the following code snippet:

```
const WASM_URL = 'wasm.wasm';
var wasm;
var mqtt;
```

3. We are using a JavaScript implementation of an MQTT client as I was not able to find an MQTT client in Go that could be built with TinyGo for the Wasm target. *Replace the values for host with the IP address from your MQTT broker.* In the future, there will surely be several clients that can be used for TinyGo Wasm projects. The code is shown in the following snippet:

```
const host = "192.2.0.23";
const port = 9001;
```

4. This function gets called when the MQTT client has successfully established a connection. When this has happened, we subscribe to the topics that are of interest to us. The code is shown in the following snippet:

```
function onConnect() {
    mqtt.subscribe("weather/data");
    mqtt.subscribe("weather/alert");
}
```

5. If we lose the connection to the MQTT broker, we want to log an error to the console. This function gets handed in later as a callback for the `connectionLost` event. The code is shown in the following snippet:

```
function onConnectionLost(responseObject) {
    if (responseObject.errorCode !== 0) {
        console.log("onConnectionLost:" +
        responseObject.errorMessage);
    }
}
```

6. When a new message arrives, we want to check what type of message we have and call the correct Go function. We determine the type of the message using the information we provide inside the message. The code is shown in the following snippet:

```
function onMessageArrived(message) {
    console.log("onMessageArrived:" +
    message.payloadString);
var payload = message.payloadString;
if (payload.indexOf("possible storm incoming") !== -1)
{
    alertHandler(payload);
} else {
    sensorDataHandler(payload);
}
}
```

7. We now add the MQTTconnect function. This function simply creates a new mqttClient and adds callback functions for the connect, connectionLost, and messageArrived events. The code is shown in the following snippet:

```
function MQTTconnect() {
    var cname = "weather-consumer";
    mqtt = new Paho.MQTT.Client(host, port, cname);
    var options = {
        timeout: 3,
        onSuccess: onConnect,
    };
    mqtt.onConnectionLost = onConnectionLost;
    mqtt.onMessageArrived = onMessageArrived;
    mqtt.connect(options);
}
```

8. Now, add the init function that is going to run our Go code, as follows:

```
function init() {
```

9. Create a new instance of go, as follows:

```
const go = new Go();
```

10. Check if the browser supports the instantiateStreaming function and, if so, load and run Wasm using this function, as follows:

```
if ('instantiateStreaming' in WebAssembly) {
WebAssembly.instantiateStreaming(fetch(WASM_URL),
    go.importObject).then(function(obj) {
        wasm = obj.instance;
        go.run(wasm);
})
}
```

11. If the browser does not support the instantiateStreaming function, we load and run Wasm using the instantiate function, as follows:

```
else {
    fetch(WASM_URL).then(resp =>
    resp.arrayBuffer()
    ).then(bytes =>
        WebAssembly.instantiate(bytes,
        go.importObject).then(function(obj) {
        wasm = obj.instance;
        go.run(wasm);
})
)
```

12. After starting our go code, we can try to connect to the MQTT broker, as follows:

```
MQTTconnect()
}
```

13. At the end of the file, add a call to the init() function, as follows:

```
init();
```

That was the complete code for our program. Now, we need to download the wasm_
exec.js file and add it to the weather-app folder. Always use the wasm_exec.js
version from your currently installed TinyGo version. You can simply download the file
for the current TinyGo release here:

https://github.com/tinygo-org/tinygo/blob/release/targets/
wasm_exec.js

In order to build and start the app, I normally use a Makefile function. The content of
the Makefile looks like this:

```
wasm-app:
 rm -rf Chapter07/html
 mkdir Chapter07/html
 tinygo build -o Chapter07/html/wasm.wasm -target wasm
 -no-debug      ch7/weather-app/wasm.go
 cp Chapter07/weather-app/wasm_exec.js ch7/html/
 cp Chapter07/weather-app/wasm.js ch7/html/
 cp Chapter07/weather-app/index.html ch7/html/
 go run Chapter07/wasm-server/main.go
```

In order to build the app and start the server, I only need to call that Makefile function
by using the following command:

```
make wasm-app
```

That works well on Linux and Mac systems, and could also work on Windows systems if
GNU Make is installed. But let's go through the process step by step so that you can also
build and run that app without using make, as follows:

1. Delete the existing html folder.

2. Create a new html folder.

3. Build the Wasm app using the Wasm target. Also, we omit debug information,
    which results in smaller binary sizes.

4. Now, copy the wasm_exec.js file to the html folder.

5. Copy the wasm.js file to the html folder.

6. Copy the index.html file to the HTML folder.

7. Run the server using the go run command.

After using the `make` command or doing the steps manually, open your browser and visit the following URL: `localhost:8080`. You should now see a site that is similar to this:

TinyGo Weather Station

Alerts:

TimeStamp Message Pressure Difference Time span

Events:

TimeStamp	Message	Temperature	Pressure	Humidity
Thu, 08 Apr 2021 18:31:00 UTC+2	sensor readings	23.4°C°C	100.54 hpa hPa	65
Thu, 08 Apr 2021 18:31:14 UTC+2	sensor readings	23.5°C°C	100.54 hpa hPa	65
Thu, 08 Apr 2021 18:31:19 UTC+2	sensor readings	23.6°C°C	100.54 hpa hPa	65
Thu, 08 Apr 2021 18:31:23 UTC+2	sensor readings	23.6°C°C	100.52 hpa hPa	64

Figure 7.13 – TinyGo weather station with opened developer tools

Excellent! We have successfully implemented a Wasm app that subscribes to topics on an MQTT broker and dynamically updates the content of a website.

Summary

In this chapter, we have learned how to use the Wi-Fi chip that is built onboard the Arduino Nano 33 IoT board. We then wrote reusable packages to use the Wi-Fi chip and the MQTT client, we discovered what MQTT is, and we learned how to publish messages to a topic. We have learned how to read sensor data from a BME280 sensor and publish this to an MQTT broker that we have set up locally.

Then, we have learned what Wasm is and implemented our first application using Wasm. We have also learned how to use a JavaScript MQTT client in order to subscribe to an MQTT topic and react to messages. While doing so, we learned how to manipulate the **Document Object Model** (**DOM**) in order to dynamically update the view.

In the next chapter, we are going to learn how to try out a Wasm app by using a login view, and will also learn how to implement bidirectional communication over MQTT.

Questions

1. What needs to be done to ensure that an MQTT message actually gets delivered?

2. Can multiple clients subscribe to the same topic on an MQTT broker?

8

Automating and Monitoring Your Home through the TinyGo Wasm Dashboard

In the previous chapter, we learned how to use the Wi-Fi chip on the Arduino Nano 33 IoT board in order to send **Message Queuing Telemetry** (**MQTT**) messages. We then consumed these messages containing weather data and weather alerts, to display them on a **WebAssembly** (**Wasm**) dashboard, but we were not able to control anything from inside the dashboard. We are now going to change this.

After working through this chapter, we will know how to secure our Wasm apps by adding a login page. We're also going to learn about security aspects when validating credentials on a client application. After building the login view, we are going to learn how to send and receive data inside a dashboard that we are going to build. By doing so, we are also going to learn some new techniques that will help us by dynamically adding and removing content. By manipulating the **Document Object Model** (**DOM**), we will know how to use bidirectional communication through MQTT. Lastly, we are going to learn about possibilities to control devices that operate at 130V (where **V** stands for **volts**) or 230V.

Knowing all this will enable us to build all sorts of home automation projects you can think of, beyond this book. In this chapter, we're going to cover the following main topics:

- Building a home automation dashboard
- Building the home automation client
- Requesting data from the microcontroller

Technical requirements

We are going to need the following components for this project:

- An Arduino Nano 33 IoT board
- A breadboard
- A **light-emitting diode** (**LED**)
- A 68 Ohm resistor
- Jumper wires

You can find the code for this chapter on GitHub at the following link: `https://github.com/PacktPublishing/Creative-DIY-Microcontroller-Projects-with-TinyGo-and-WebAssembly/tree/master/Chapter08`

The Code in Action video for the chapter can be found here: `https://bit.ly/3uPLI7X`

Building a home automation dashboard

After you have finished this book, you might want to build lots of cool projects that could include LED stripes or a motion-sensor-controlled light, or you might add a motor to your curtains to open or close them based on the light intensity or the time. These would be really cool projects, but now imagine that you're sitting on the couch and want to watch a movie, but the sun is too bright and did not exceed the threshold to start the motor that controls the curtains. What can we do in such a situation? Do we stand up and close the curtains manually, or do we open a Wasm app on our smartphone or tablet in order to control the motor for the curtains by just pressing a button on an app? You might also want to check if the LED stripe in the living room is still turned on, but you do not want to get out of bed to check. In that case, it would be great to have a dashboard that informs you about its status. In this section, we are going to build a Wasm app that provides a login page where the user is able to enter a username and a password before they can log in. The page should then be transitioned to a dashboard that provides functionality to enable or disable the lights in a specific room.

We will start with a reusable MQTT JavaScript component that can be used until a TinyGo-compatible MQTT library has been created.

Creating a reusable MQTT component

In *Chapter 7, Displaying Weather Alerts on the TinyGo Wasm Dashboard*, we embedded the MQTT client into a wasm.js file. This worked well for the project but is not reusable. Therefore, we are now going to create a reusable component.

To do so, start off by creating a new folder named Chapter08 for this project. Inside the newly created folder, create a new folder called light-control. This new folder is going to contain all the files that are needed for the Wasm app.

Now, create a new file inside the light-control folder and name the file mqtt. js. The project structure should now look like this:

Figure 8.1 – Project structure

Inside the `mqtt.js` file, follow these steps to implement it:

1. First, we define a variable that holds the MQTT client and constants for the MQTT broker. We also use `strict` mode again, to prevent us from using undefined variables. Strict mode also eliminates some silent errors and exchanges them to throw errors instead, and enables JavaScript engines to perform optimizations that were otherwise not possible. Using `strict` mode could lead to faster execution. The `host` and `port` values *need to be set to your own MQTT broker* host and port if the broker is not running locally. The code is shown in the following snippet:

```
'use strict';
var mqtt;
const host = "192.2.0.23";
const port = 9001;
const cname = "home-automation-dashboard";
```

2. We then add a function that simply logs to the console when the connection to the MQTT broker has been successfully established, as follows:

```
function onConnect() {
    console.log("Successfully connected to mqtt broker");
}
```

3. As the Wasm app is being executed on the client, it is possible that we will lose the connection to the MQTT broker. This could be caused by an unstable Wi-Fi connection. If that happens, we want to attempt to create a new connection. We can do this by running the following code:

```
function onConnectionLost(err) {
    if (err.errorCode !== 0) {
        console.log("onConnectionLost:" +
            err.errorMessage);
    }
    MQTTconnect();
}
```

4. We now need to add a callback for the `messageArrived` event. When a new message arrives, we want to call a message handler that is exported by the Go code. This works as follows:

```
function onMessageArrived(message) {
    console.log(
        "onMessageArrived:" + message.payloadString);
    handleMessage(message.payloadString);
}
```

5. We next want to be able to publish messages. In this case, we set the **quality of service (QOS)** to 1 as we want to make sure that a message is definitely received by consumers. Furthermore, we do not need to retain messages. In later projects, you could also parameterize the QOS level and the `retain` flag. The code is shown in the following snippet:

```
function publish(topic, message) {
    mqtt.send(topic, message, 1, false);
}
```

6. Establish the connection to the MQTT broker, as follows:

```
function MQTTconnect() {
    console.log("mqtt client: connecting to " + host +
        ":" + port);

    mqtt = new Paho.MQTT.Client(host, port, cname);
    var options = {
        timeout: 3,
        onSuccess: onConnect,
    };

    mqtt.onConnectionLost = onConnectionLost;
    mqtt.onMessageArrived = onMessageArrived;

    mqtt.connect(options);
}
```

This is everything we need for our reusable MQTT component. All we need to do when integrating it into projects is this:

1. Expose a `handleMessage()` function in the Go code.

2. Set the `hostname, port, and cname` values to the MQTT broker in the JavaScript file.

The next step is to set up the so-called glue code that connects the JavaScript code with the Go code.

Setting up the Wasm instantiation code

The Wasm instantiation code is nearly the same every time. It only changes if we want to add some project-specific code in it. So, let's quickly create a new file named `wasm.js` inside the `light-control` folder. Now, run the following standard code to initialize a Wasm app inside the new file:

```
'use strict';
const WASM_URL = 'wasm.wasm';
var wasm;

function init() {
    const go = new Go();
    if ('instantiateStreaming' in WebAssembly) {
        WebAssembly.instantiateStreaming(fetch(WASM_URL),
                go.importObject).then(function (obj) {
            wasm = obj.instance;
            go.run(wasm);
        })
    } else {
        fetch(WASM_URL).then(resp =>
            resp.arrayBuffer()
        ).then(bytes =>
            WebAssembly.instantiate(bytes,
                    go.importObject).then(function (obj) {
                wasm = obj.instance;
                go.run(wasm);
            })
        )
```

```
    }
}
init();
```

This is nearly the same code as in *Chapter 7, Displaying Weather Alerts on the TinyGo Wasm Dashboard*, in the *Implementing the weather app* section, but this time we did not include the MQTT client code inside the file. You can use this file for every project beyond this book.

The next step is to add the `wasm_exec.js` file. We can either download it from the TinyGo GitHub repository or copy it from our local installation. On Unix-based systems, you can use the following command to copy the file:

```
cp /usr/local/tinygo/targets/wasm_exec.js /path/to/Chapter08/
light-control
```

The path to the `wasm_exec.js` file is different on Windows. When using the preceding command, you need to insert your own path to the TinyGo installation. The path basically follows this pattern:

```
/path/to/your/tinygo/installation/target/wasm_exec.js
```

That's everything we need in terms of JavaScript code. We can now go on to create our **HyperText Markup Language** (**HTML**) template file.

Creating the HTML template

In *Chapter 7, Displaying Weather Alerts on the TinyGo Wasm Dashboard*, in the *Implementing the weather app* section, we defined our base structure inside the HTML file, but this time our HTML template is going to be much shorter. We are only going to include the needed header and define an empty body element as we are going to create all HTML elements dynamically using DOM manipulation from inside the Go code.

To do so, create a new file named `index.html` inside the `light-control` folder. The body element needs to get an `id` value as we are going to identify the element using the ID. We also import all needed JavaScript files in the header. This is what it will look like:

```
<!DOCTYPE html>
<html>

<head>
    <meta charset="utf-8" />
    <title>TinyGo Home Automation</title>
```

```
    <meta name="viewport" content="width=device-width,
        initial-scale=1" />
    <script src="wasm_exec.js"
        type="text/javascript"></script>
    <script src="wasm.js" type="text/javascript"></script>
    <script src="mqtt.js" type="text/javascript"></script>
    <script src="https://cdnjs.cloudflare.com/ajax/libs/paho-
        mqtt/1.0.1/mqttws31.min.js"
        type="text/javascript"></script>
</head>


<body id="body-component"></body>


</html>
```

That's everything we need for the HTML template. The next step is writing the login view.

Implementing the login view logic

The login component needs to add the login view to the HTML document and also implement logic to handle the user input. Let's create a new folder named login inside the light-control folder and create a new file named userinfo.go inside the newly created folder.

The userinfo.go file simply holds the UserInfo element, which looks like this:

```
type UserInfo struct {
    LoggedIn bool
    UserName string
    LoggedInAt time.Time
}
```

Now, we create a new login.go file inside the login folder and implement the view by following these steps:

1. We need username and password values for the login, so we define them as follows:

    ```
    const user = "tinygo"
    const password = "secure1234"
    ```

2. We only need to fetch the document a single time, so we just store it inside a package-level variable, as follows:

```
var doc = tinydom.GetDocument()
```

3. Now, we define a service that only needs to hold a channel. The channel is later being used to propagate the logged-in username to the other components. The code is shown in the following snippet:

```
type Service struct {
    channel chan string
}
```

4. We define a constructor function that accepts a channel and returns a new instance of Service, as follows:

```
func NewService(channel chan string) *Service {
    return &Service{channel: channel}
}
```

5. The next step is to implement the logic to create the view. We want to simulate having a multipage app by telling the browser to change the **Uniform Resource Locator** (**URL**) by pushing a new state. The code to do this is shown in the following snippet:

```
func (service *Service) RenderLogin() {
    tinydom.GetWindow().PushState(nil, "login", "/login")
```

6. We now create a new div tag that is going to hold all subsequent elements, as follows:

```
div := doc.CreateElement("div").
    SetId("login-component")
```

7. We then set an h1 that also tells the user the name of the component, as follows:

```
h1 := doc.CreateElement("h1").
    SetInnerHTML("Login")
```

8. Now, we create a form that holds the input fields. So, we simply create a new instance of `form` and also create a new `userName` input field with the corresponding label, which works as follows:

```
loginForm := form.New()
userNameLabel := label.
        New().
        SetFor("userName").
        SetInnerHTML("UserName:")

userName := input.
        New(input.TextInput).
        SetId("userName")
```

9. We now want to add an input field of type `password` that obscures the input. To do this, run the following code:

```
passwordLabel := label.
        New().
        SetFor("password").
        SetInnerHTML("Password:")

password := input.
        New(input.PasswordInput).
        SetId("password")
```

10. As we now have both input fields, we need a button that emits `click` and `keyPress` events that we can use to trigger the `login` logic. Here is the code we run to do this:

```
login := input.New(input.ButtonInput).
        SetValue("login").
        AddEventListener("click",
            js.FuncOf(service.onClick)).
        AddEventListener("keypress",
            js.FuncOf(service.onKeyPress))
```

11. We have now created all components that we need inside our `loginForm`, so we can go on and append them to `loginForm`, as follows:

```
loginForm.AppendChildrenBr(
        userNameLabel,
        userName,
        passwordLabel,
        password,
        login,
)
```

12. The last thing to do is to append the previous element to the `div`. We append everything inside a `div` so that we can easily delete the elements again. In order to display the newly created elements, we just append the `div` in the body, as follows:

```
div.AppendChildren(h1, loginForm.Element)
body := doc.GetElementById("body-component")
body.AppendChild(div)
}
```

We can now create the view itself. The only thing missing here is the logic to handle `EventListener` from the `login` button, as well as the login logic itself. To do so, follow these last few steps for this component:

1. When the user clicks on the `login` button, we simply want to attempt a login. The following code snippet illustrates this:

```
func (service *Service) onClick(this js.Value, args
        []js.Value) interface{} {
    service.login()

    return nil
}
```

2. When the input button is focused and the user hits the *Enter* button, we also want to attempt a login. We wrap the event `args` into a `tinydom` event that provides us with convenience functions, as illustrated in the following code snippet:

```
func (service *Service) onKeyPress(this js.Value, args
        []js.Value) interface{} {
```

```
    if len(args) != 1 {
        println("bad number of arguments in keyPress
            event")

        return nil

    }

    event := tinydom.Event{Value: args[0]}
    if event.Key() == "Enter" {
        service.login()
    }

    return nil
}
```

3. The `login` function fetches the input from the `username` and `password` input fields and compares them to our defined credentials. When invalid credentials are found, we trigger an alert. The most important bit in this function is the need to wrap the call that writes into the channel inside a goroutine. If we did not wrap a goroutine around it, the code could not compile. Refer to the following code:

```
func (service *Service) login() {
    userElem := input.FromElement(
        doc.GetElementById("userName"))
    userName := userElem.Value()

    if userName != user {
        tinydom.GetWindow().Alert("Invalid username or
            password")
        return
    }

    passwordElem := input.FromElement(
        doc.GetElementById("password"))
    passwordInput := passwordElem.Value()

    if passwordInput != password {
```

```
        tinydom.GetWindow().Alert("Invalid username or
            password")
        return
    }


    go func() {
        service.channel <- userName
    }()
}
```

Excellent! We have completed the login component. But what would this view look like in a browser? Let's check the following screenshot to find out:

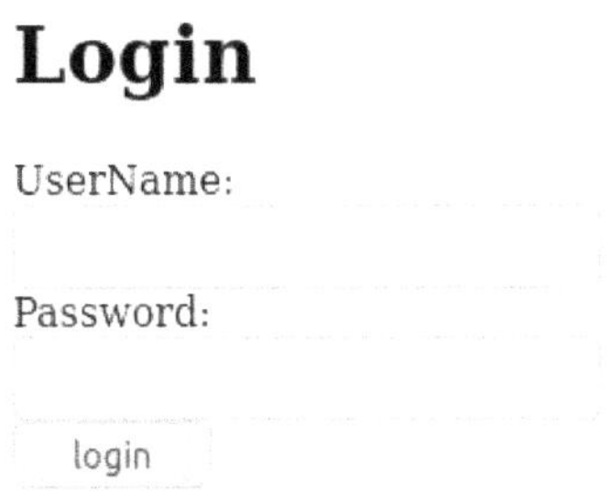

Login

UserName:

Password:

login

Figure 8.2 – Login view rendered by a browser

Before we implement the dashboard component, we have to talk about some security aspects of this login component. We have defined the credentials that are needed to log in inside the code. This code is going to be downloaded by the browser in the form of a Wasm binary file. As we provide the code that includes the credentials to the client, this procedure is considered unsecure. Let's have a look at the following screenshot that shows a part of the Wasm binary:

`userNametinygopasswordsecure1234`

Figure 8.3 – Credentials leaked in binary file

To find the credentials inside the Wasm binary, I simply opened the binary in a text editor and searched for the password. So, what other possibilities do we have to keep credentials safe? Here are a couple of options:

- Make an HTTP call to a **REpresentational State Transfer (REST) application programming interface (API)** that validates credentials.

- Use any JavaScript library that is able to talk to an **Open Authorization 2 (OAuth 2)** service.

There might be plenty of other possibilities, but they all boil down to moving the actual credential validation logic to any kind of external API. But for our scope, this solution is good enough to validate the credentials inside the client. The next step is to implement the dashboard component.

Implementing the dashboard component

We are now going to implement our home automation dashboard. The dashboard is going to hold a list of components with associated actions that are represented by buttons. We also want to log out the user after a 5-minute period of inactivity. Before we dive into the code, we need to create a new folder named `dashboard` inside the `light-control` folder and create a new `dashboard.go` file inside. Now, follow these steps to implement the logic:

1. We save a reference to the current document, as follows:

    ```
    var doc = tinydom.GetDocument()
    ```

2. The service object holds a channel that we use to signal a logout. The `UserInfo` object will later be used to check `loginTime`, which will be used as an inactivity timer. We also get `UserName` from `UserInfo`. The code is shown in the following snippet:

    ```
    type Service struct {
        user login.UserInfo
        logoutChannel chan struct{
    }
    ```

    The constructor function needs to get the channel injected as we react to logout events that are being sent to the channel from the `wasm.go` file. The code is shown in the following snippet:

    ```
    func New(logout chan struct{}) *Service {
        return &Service{
            logoutChannel: logout,
        }
    }
    ```

3. We want to be able to trigger a connection attempt to the MQTT broker from inside the Go code, so we call the `js` function that resides inside the `mqtt.js` file, as follows:

```
func (service *Service) ConnectMQTT() {
    println("connecting to mqtt")
    js.Global().
        Get("MQTTconnect").
        Invoke()
}
```

4. Now, we define a function that can be used as a callback to an eventListener. As this function is being called from within JavaScript, we need to fulfill a function signature that takes `js.Value` and a `[]js.Value` parameters and returns `interface{}`, as follows:

```
func (service *Service) logout(this js.Value, args
        []js.Value) interface{} {
    service.logoutChannel <- struct{}{}
    return nil
}
```

5. The `bedroomOn` function is being used as callback for the JavaScript code and will be called when the user clicks the On button. The code is shown in the following snippet:

```
func (service *Service) bedroomOn(this js.Value, args
        []js.Value) interface{} {
```

6. When the user performs any action, we need to check if the activity timer timed out. We do so by checking the `loggedInAt` timestamp. If the user is inactive for more than 5 minutes we perform a logout, as follows:

```
if time.Now().After(service.user.LoggedInAt.Add(5 *
        time.Minute)) {
    println("timeOut: perform logout")
    service.logout(js.ValueOf(nil), nil
    )return nil
}
```

7. Now, we simply need to invoke the `publish` function in the JavaScript code and reset the `loggedInAt` timer, as follows:

```
println("turning lights on")
// room # module # action
js.Global().Get("publish").Invoke("home/bedroom/lights",
    "on")

service.user.LoggedInAt = time.Now()
return nil
```

8. Turning the lights off works in a similar way to turning them on. The only difference is the payload of the message. We send `off` instead of `on` here, which looks like this:

```
func (service *Service) bedroomOff(this js.Value, args
    []js.Value) interface{} {

    if time.Now().After(service.user.LoggedInAt.Add(5 *
            time.Minute)) {
        println("timeOut: perform logout")
        service.logout(js.ValueOf(nil), nil)
        return nil

    }

    println("turning lights off")
    js.Global().Get("publish").Invoke("home/bedroom
        /lights","off")
    service.user.LoggedInAt = time.Now()
    return nil
}
```

We have successfully implemented the complete control logic. Now, we need to implement the logic that creates the view. These are the necessary steps to do this:

1. When we create the dashboard view, we have information about which user just logged in, so we store this as follows:

    ```
    func (service *Service) RenderDashboard(user
            login.UserInfo) {
        service.user = user
    ```

2. Just as in the login view, we tell the browser to display another URL by pushing a new state, as follows:

    ```
    tinydom.GetWindow().
        PushState(nil, "dashboard", "/dashboard")
    ```

3. We create a new `div` element and set an `Id` value so that we can identify the element later on, to remove it when logging out. The code is shown in the following snippet:

    ```
    body := doc.GetElementById("body-component")
    div := doc.CreateElement("div").
            SetId("dashboard-component")
    ```

4. We now greet the user by their name, as follows:

    ```
    h1 := doc.CreateElement("h1").
            SetInnerHTML("Dashboard")
    h2 := doc.CreateElement("h2").
            SetInnerHTML(fmt.Sprintf("Hello %s",
                service.user.UserName))
    ```

5. As we want to have a way to easily add new components to the dashboard, we use a table to control the components. That way, we can simply add new table rows later on. Of course, we could also create new custom components or use any other kind of structure, but adding rows to a table is easy to understand. The whole process looks like this:

    ```
    tableElement := table.New().
        SetHeader("Component", "Actions")
    tbody := doc.CreateElement("tbody")

    tr := doc.CreateElement("tr")
    ```

```
componentNameElement := doc.CreateElement("td").
    SetInnerHTML("Bedroom Lights")
componentControlElement := doc.CreateElement("td")

onButton := input.New(input.ButtonInput).
    SetValue("On").
    AddEventListener("click",
        js.FuncOf(service.bedroomOn))
offButton := input.New(input.ButtonInput).
    SetValue("Off").
    AddEventListener("click",
        js.FuncOf(service.bedroomOff))

componentControlElement.AppendChildren(onButton,
    offButton)
tr.AppendChildren(componentNameElement,
    componentControlElement)
tbody.AppendChildren(tr)


tableElement.SetBody(tbody)
```

6. Besides a logout based on inactivity, we want the user to have the possibility to log out manually. Here's how to set this up:

```
logout := input.New(input.ButtonInput).
    SetValue("logout").
    AddEventListener("click", js.FuncOf(service.logout),
)
```

7. The last steps append all child elements to the div and append the div to the body, as illustrated in the following code snippet:

```
div.AppendChildren(
    h1,
    h2,
    tableElement.Element,
    tinydom.GetDocument().CreateElement("br"),
    logout,
```

```
    )
body.AppendChild(div)
    }
```

Great! We have now fully implemented the logic that is needed to create the view. When rendered by a browser, the view looks similar to this:

Dashboard

Hello tinygo

Component	Actions	
Bedroom Lights	On	Off

logout

Figure 8.4 – Dashboard view

Now, we only need to implement the main logic before the app itself is complete.

Implementing the main logic

We are going to split the logic for the different components (login, dashboard) into separate files. The wasm.go file, which we now create inside the light-control folder, is going to hold the main() function and is used to control the flow through the app.

We are now going to introduce a new library called tinydom. The tinydom library wraps the syscall/js API and also provides additional data types such as Video, Form, or Label. Using this library, we can save lots of **lines of code** (**LOC**). As tinydom works on js.Value types internally, it is fully interoperable with the syscall/js API. You can install tinydom using the following command:

```
go get github.com/Nerzal/tinydom
```

As this is now set up, let's go on to implement the logic by following these steps:

1. Above the `main` function, we define some variables. We define them outside of the `main` function as we are going to need them inside the functions. The code is shown in the following snippet:

    ```
    var window = tinydom.GetWindow()
    var loginService *login.Service
    var loginState login.UserInfo
    var dashboardService dashboard.Service
    ```

2. We use the `main` function to render the login screen, as well as setting up the login and logout event handler. This is done as follows:

    ```
    func main() {
        loginState = login.UserInfo{}
        loginChannel := make(chan string, 1)
        loginService = login.NewService(loginChannel)
        loginService.RenderLogin()
        go onLogin(loginChannel)

        logoutChannel := make(chan  struct{}, 1)
        go onLogout(logoutChannel)

        dashboardService = dashboard.New(logoutChannel)
        wait := make(chan struct{}, 0)
        <-wait
    }
    ```

3. When receiving a login event from the channel, we initialize `loginState`, connect to MQTT, and render the dashboard view, as follows:

    ```
    func onLogin(channel chan string) {
        for {
            userName := <-channel
            println(userName, "logged in!")

            loginState.UserName = username
            loginState.LoggedIn = true
    ```

```
            loginState.LoggedInAt = time.Now()
            removeLoginComponent()
            dashboardService.ConnectMQTT()
            dashboardService.RenderDashboard(loginState)
        }
    }
```

4. In order to remove an object from the view, we simply remove it from the DOM. We do so by fetching the body element and removing the child with the `login-component` ID, as follows:

```
func removeLoginComponent() {
    doc := tinydom.GetDocument()
    doc.GetElementById("body-component").
        RemoveChild(doc.GetElementById(
            "login-component"))
}
```

5. We also want to be able to remove the dashboard view to be able to go back to the login view. We do this by running the following code:

```
func removeDashboardComponent() {
    doc := tinydom.GetDocument()
    doc.GetElementById("body-component").
        RemoveChild(doc.GetElementById(
            "dashboard-component"))
}
```

6. When we receive a logout event from the channel, we remove the dashboard view, reset the login state, and render the login view again, as follows:

```
func onLogout(channel chan struct{}) {
    for {
        <-channel
        println("handling logout event")
        removeDashboardComponent()
        loginState = login.UserInfo{}

        loginService.RenderLogin()
```

```
        }
    }
```

That's everything we need for our main logic. The next step is to implement a server that serves the app to clients.

Serving the app

Serving the app works in a similar way to serving the app in *Chapter 7*, *Displaying Weather Alerts on the TinyGo Wasm Dashboard*, but we add an extra trick in here. When the user refreshes the page or tries to visit one of the URLs that we set by pushing a state, the server would normally be unaware of these URLs. That's why we redirect the client to the correct URL. We handle that case by simply redirecting the user to the root URL.

Now, add the following code into a main.go file that resides in a newly created folder named wasm-server, which is inside the Chapter08 folder:

```go
const dir = "Chapter08/html"
var fs = http.FileServer(http.Dir(dir))

func main() {
    log.Print("Serving " + dir + " on http://localhost:8080")
    http.ListenAndServe(":8080",
        http.HandlerFunc(handleRequest))
}

func handleRequest(
        resp http.ResponseWriter, req *http.Request) {

    resp.Header().Add("Cache-Control", "no-cache")
    if strings.HasSuffix(req.URL.Path, ".wasm") {

        resp.Header().Set("content-type", "application/wasm")
    }
    requestURI := req.URL.RequestURI()

    if strings.Contains(requestURI, "dashboard") ||
```

```
        strings.Contains(requestURI, "login") {

        http.Redirect(resp, req, "http://localhost:8080",
            http.StatusMovedPermanently)
        return
    }
fs.ServeHTTP(resp, req)
}
```

We have completed the app and the server that serves the app. Let's now build and run everything. We will be using a **Makefile** for this example, but you could also use a Docker container, a shell script, or something similar. We need to build the Wasm app, copy all dependencies, and start the server. The Makefile approach looks like this:

```
light-control:
rm -rf Chapter08/html
mkdir Chapter08/html
tinygo build -o Chapter08/html/wasm.wasm -target wasm -no-debug
Chapter08/light-control/wasm.go
cp Chapter08/light-control/wasm_exec.js Chapter08/html/
cp Chapter08/light-control/wasm.js Chapter08/html/
cp Chapter08/light-control/mqtt.js Chapter08/html/
cp Chapter08/light-control/index.html Chapter08/html/
go run Chapter08/wasm-server/main.go
```

In order to run the server, we use the following command:

```
make light-control
```

When this is successful, go on and try out our app by visiting the following URL in a browser:

```
localhost:8080
```

When using the Mosquitto Docker container, don't forget to check that the container has been started and that the container does not run. Simply start it by using the following command:

```
docker start mosquitto
```

As we have now successfully built a Wasm app that is able to publish messages to an MQTT broker, we can now go on and create a client that consumes these messages, and this is exactly what we are going to do in the next section.

Building the home automation client

Home automation basically consists of *activating and deactivating things* based on a *precondition*. For example, we might want to turn on a light when someone enters a room at night. Throughout this book, we have activated and deactivated many things based on preconditions, but most of them were not connected to a network. We are now going to learn how we can send signals over a network. These signals are going to be used as preconditions. After completing this section, we will be fully prepared for building our own home automation clients that can be triggered over a network.

The client that is going to run on the Arduino Nano 33 IoT is simply going to connect to an MQTT broker and then subscribe to a topic. When a message comes in for the topic, we need to deserialize the message and perform the action that is defined in the message.

For our example project, we are going to turn an LED on and off. Of course, a single LED might not be enough to light up a complete bedroom, so we will talk about other real-world solutions at the end of this section. Let's start by setting up the circuit.

Setting up the circuit

The circuit for this project is fairly simple. Just follow these steps to set everything up:

1. Place an LED with the cathode in *E40* on the breadboard.

2. Connect A41 (*GND*) with the *GND* lane on the power bus.

3. Connect the anode of the LED with pin D4 and place a 68 Ohm resistor in between. If you don't have a 68 Ohm resistor, you can also use a 100 Ohm one. Connect *B52* with the *GND* lane on the power bus.

The result should look similar to this:

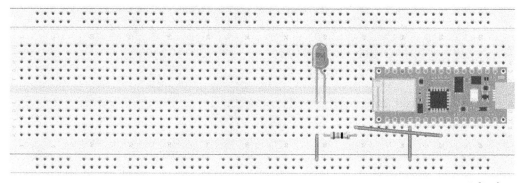

fritzing

Figure 8.5 – Light control circuit (image is taken from Fritzing)

If you are unsure which technical specs your LEDs have because you simply do not have a datasheet, have a look at the following URL. This provides a resistor calculator, as well as good voltages for different LED colors:

```
https://www.digikey.de/en/resources/conversion-calculators/
conversion-calculator-led-series-resistor
```

Great! We are now all set up and ready to implement the logic.

Implementing the logic

For our final project, we need to create a new folder named `light-control-client` inside the `Chapter08` folder and create a new `main.go` file inside. The logic of the `main` function is just used to initialize everything, while the actual logic is going to reside in separate functions. To implement it, follow these steps:

1. Above the main function, we add constants for the Wi-Fi credentials and the LED pin. We simply need to replace the SSID and password with our own data, which looks like this:

    ```
    const ssid = ""
    const password = ""
    const bedroomLight = machine.D4
    ```

2. Now, inside the main function, we want to control the LED. To do so, we need to configure the pin as output, as shown in the following code snippet:

```
time.Sleep(5 * time.Second)
bedroomLight.Configure(machine.PinConfig{Mode:
    machine.PinOutput})
```

3. The next step is to establish the Wi-Fi connection, as seen in the following code snippet:

```
wifiClient := wifi.New(ssid, password)

println("configuring nina wifi chip")
err := wifiClient.Configure()
if err != nil {
    printError("could not configure wifi client", err)
}
println("checking firmware")
wifiClient.CheckHardware()
wifiClient.ConnectWifi()
```

4. Now, we need to connect to the MQTT broker. You need to replace the IP address with the IP address of your MQTT broker, as follows:

```
mqttClient := mqttclient.New("tcp://192.168.2.102:1883",
    "lightControl")
println("connecting to mqtt broker")
err = mqttClient.ConnectBroker()
if err != nil {
    printError("could not configure mqtt", err)
}
println("connected to mqtt broker")
```

5. In order to subscribe to a topic, we need to hand in the QOS level and a function that is called when a message on that topic arrives, which looks like this:

```
err = mqttClient.Subscribe(
        "home/bedroom/lights",
        0,
        HandleActionMessage,
)
if err != nil {
printError("could not subscribe to topic", err)
}
```

6. The last step is to add in a blocking function so that the program does not terminate, which can be seen in the following code snippet:

```
println("subscribed to topic, waiting for messages")
select {}
```

That's all we need to initialize everything. We now only need to implement the logic that handles the incoming messages. To do so, follow these steps:

1. First, we need to deserialize the incoming message by splitting the string, and then call functions depending on the room that is being delivered. If we receive an invalid message or finish handling the message, we Ack the message, as follows:

```
func HandleActionMessage(client mqtt.Client, message
        mqtt.Message) {
    println("handling incoming message")
    payload := string(message.Payload())

    controlBedroom(client, payload)
    message.Ack()

}
```

2. In the next step, we simply execute the correct functions based on the module and action provided. The complete function is implemented in the following snippet:

```
func controlBedroom(module, action string) {
    switch action {
    case "on":
        controlBedroomlights(client, true)
    case "off":
        controlBedroomlights(client, false)
    default:
        println("unknown action:", action)
    }
}
```

3. Now, we just need to activate or deactivate the LED, as follows:

```
func controlBedroomlights(action bool) {
    if action {
        bedroomLight.High()
    } else {

        bedroomLight.Low()
    }
}
```

4. We want to stop the execution and print the error repeated, while initializing everything. For that case, we use the following helper function:

```
func printError(message string, err error) {
    for {

        println(message, err.Error())
        time.Sleep(time.Second)
    }
}
```

That's everything we need to implement the client. We can now go on and flash the program, using the following command:

```
tinygo flash --target arduino-nano33 Chapter08/light-control-
client/main.go
```

While the program is running, we are now able to turn the LED on and off using the Wasm app. So, now go on and try that.

Okay—you tried that; everything is working as expected and now you want to know what's next. What if something went wrong and the LED never activates or deactivates?

In such a case, I highly recommend watching the output of the serial port in PuTTY. If everything looks fine there, the next thing you can try is to send MQTT messages to the broker by using MQTT Explorer. If you still have no luck, you should double-check your wiring; and if nothing else helps, you might want to try to directly flash the code from the GitHub repository.

Now that everything is working as intended, you might think that *only* being able to activate and deactivate the light is nice, but what about displaying the current status of the light on the dashboard? Let's do this as a next step.

Requesting data from the microcontroller

We might want to know if the light is currently turned on or off inside the living room, without having to walk all the way into the room. So, it would be great if the Wasm app *could request the status of the light* and display it.

Now, let's imagine we have one or multiple microcontrollers in different rooms, listening to messages. For this example, we do not want the microcontroller to continuously report the state of the light as this would cause unnecessary network traffic. So, we go on and send a message to request the data. The microcontrollers are subscribed to the status topic and get the message delivered. After receiving the status request, they answer it by each sending a status message.

This process is represented in the following diagram:

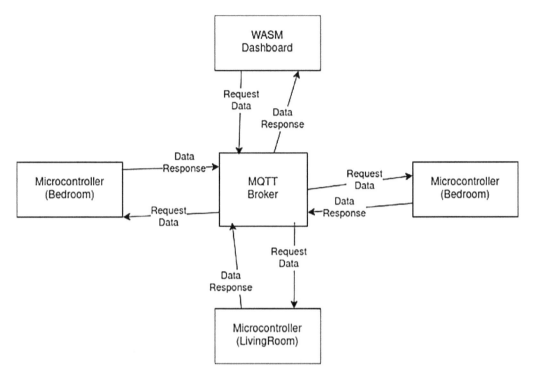

Figure 8.6 – Architecture diagram

In order to implement that behavior, one microcontroller is sufficient. So, let's go on and update our code accordingly. To do so, follow these steps:

1. Inside the wasm.js file, we subscribe to the home/status topic. This is the topic in which the microcontrollers are going to publish status messages. We also want to call a go function when the connection has been established. Refer to the following code:

```
function onConnect() {
    console.log("Successfully connected to mqtt broker");
    mqtt.subscribe("home/status")
    handleOnConnect()
}
```

2. Inside the `dashboard.go` file, we add a `Boolean` to save the status of the
    bedroom lights inside the `Service` struct, as follows:

```go
type Service struct {
    user login.UserInfo
    bedroomLights bool
    logoutChannel chan bool
}
```

3. We need to expose the `handleMessage` function to the JavaScript code so that
    it can be called when a new message arrives. We also expose a `handleConnect`
    function to the JavaScript code, which is called when the connection to the broker
    has been established. The code is shown in the following snippet:

```go
func New(logout chan bool) Service {
    js.Global().
    Set("handleMessage", js.FuncOf(handleMessage))
    js.Global().
    Set("handleOnConnect", js.FuncOf(handleOnConnect))
    return Service{
        logoutChannel: logout,
    }
}
```

4. As we want to add a new column to the table, we need to add a new column header.
    We can do this with the following code:

```go
tableElement := table.New().
    SetHeader(
        "Component",
        "Actions",
        "Status",
    )
```

5. We now want to add a new `Status` column inside the table, so we need to
    add some lines of code inside the `RenderDashboard` function. Right below
    `controlElement`, we add a new `statusElement`, as follows:

```go
componentControlElement := doc.CreateElement("td")
statusElement := doc.CreateElement("td").
```

```
SetId("bedroom-light-status").
SetInnerHTML("off")
```

6. As we have added a column, we need to add it to the table row. We can do this by running the following code:

```
tr.AppendChildren(
    componentNameElement,
    componentControlElement,
    statusElement,
)
```

7. Now, we add a new function that lets us request the status. We use the home/status-request topic for that purpose. This is illustrated in the following code snippet:

```
func requestStatus() {
    js.Global().
    Get("publish").
    Invoke("home/status-request", "")
}
```

8. As we now have the ability to request the status, we just need to invoke it to get the status updates. We do it right after the MQTT connection has been established, as follows:

```
func handleOnConnect(this js.Value, args []js.Value)
        interface{} {
    requestStatus()
    return nil
}
```

9. The last thing that we need to add is handling the message. So, let's split the message into room, component, and action and call the correct function depending on room and component, as follows:

```
func handleMessage(this js.Value, args []js.Value)
interface{} {
    message := args[0].String()
```

```
println("status message arrived:", message)
messageParts := strings.Split(message, "#")

room := messageParts[0]
component := messageParts[1]

switch room {
    case "bedroom":
        switch component {
            case "lights":
                doc.GetElementById("bedroom-light-
                    status").
                SetInnerHTML(messageParts[2])default:
                    println("unknown component:",
                        component)}
        default:
            println("unknown room:", room)}
    return nil
}
```

We have successfully added all we need to the Wasm app. Let's now extend the logic of the `light-control-client` program. To do so, follow these steps:

1. We need to save the current status of the lights, so we add a new variable at the package level, as follows:

    ```
    var bedroomLightStatus = false
    ```

2. In the `main` function, we subscribe to the `home/status-request` topic, as illustrated in the following code snippet:

    ```
    err = mqttClient.Subscribe("home/status-request", 0,
        HandleStatusRequestMessage)
    if err != nil {

        printError("could not subsribe to topic", err)
    }
    ```

3. We now need to implement the handler for the status request. We simply report the status and `Ack` the message afterward, as illustrated in the following code snippet:

```go
func HandleStatusRequestMessage(client mqtt.Client,
        message mqtt.Message) {
    reportStatus(client)
    message.Ack()
}
```

4. The `reportStatus` function just needs to check and report the status. This can be done by running the following code:

```go
func reportStatus(client mqtt.Client) {
    status := "off"
    if bedroomLightStatus {
        status = "on"
    }

    token := client.Publish(
            "home/status",
            0,
            false,
            fmt.Sprintf("bedroom#lights#%s", status),
    )

    if token.Wait() && token.Error() != nil {
        println(token.Error())
    }
}
```

5. Inside the `HandleActionMessage` function, we need to pass `mqtt.Client` as an additional parameter to the `controlBedroom` function. We can do this by running the following code:

```go
controlBedroom(
    client,
    splittedString[1],
    splittedString[2],
)
```

6. We now also need to add `mqtt.Client` to the `controlBedroom` parameter list. We can do this by running the following code:

```
func controlBedroom(client mqtt.Client, module, action
     string) {
```

7. We then pass the client in to the `controlBedroomlights` function, as follows:

```
controlBedroomlights(client, true)
```

8. The last step is to update and report the status in the `controlBedroomLights` function. We also update the status here so that we get feedback in the Wasm app after clicking on the **on/off** buttons. The code for this is shown in the following snippet:

```
func controlBedroomlights(client mqtt.Client, action
     bool) {
    if action {
        bedroomLight.High()
        bedroomLightStatus = true
    } else {
        bedroomLight.Low()
        bedroomLightStatus = false
    }

    reportStatus(client)
}
}
```

Excellent! The client can now check the status of the lights inside the Wasm app.

Well, congratulations! You have finished all of the projects in this book. Let's now have a look at possible alternative solutions to our current implementation.

Checking other implementation ideas

Lighting up a small LED by pressing a button on a Wasm app is exciting but does not really help in terms of home automation. An LED can be considered as a placeholder for literally anything you can think of. We have implemented the logic to trigger any kind of action. What possibilities do we have to control real lights or other components?

Using smart sockets

One option is to use smart sockets, which are controllable using Wi-Fi or Bluetooth. Most of them do not provide an open API and require you to reverse-engineer the signals to control them, but there are also some manufacturers that provide API references for their products.

An example of this is the NETIO PowerBOX 3Px, which is a socket that supports lots of APIs such as MQTT, HTTP, **JavaScript Object Notation** (**JSON**), and **Transmission Control Protocol** (**TCP**), among others. Another example is WIFIPLUG—they also produce smart sockets that have open APIs available.

Using a relais

We have learned how to control a relais when building our automatic plant-watering system. Some relais and boards support voltages up to 230V and 10 **amps** (**A**), which is sufficient to power nearly any electrical device. Although the relais might be able to handle currencies of 230V or 130V, you should never work mains-voltage. Lots of nice projects can be built with currents up to 12V.

Using TLS

When developing **Internet of Things** (**IoT**) applications, it's important to consider security. At the time of writing, the Wi-Fi driver implementation of the Arduino Nano 33 IoT does not support TLS. This is a topic that is actively being worked on and will definitely be implemented soon. So, when implementing functionalities that operate outside of your local network, you should definitely aim to use TLS. Also, as mentioned, when implementing the login view, we learned that embedding credentials into Wasm is not as secure as embedding credentials into the binary file.

We have now learned that there are several manufacturers for smart sockets that have an open API, which makes it easy to integrate them into our projects **safely**. We have also learned that we can make use of a relais to control LED stripes or other devices.

Summary

In this chapter, we have learned how to build a Wasm app that creates its views fully and dynamically. We have learned this by manipulating the DOM. We have also learned how to handle user input in Wasm and how to create reusable JavaScript components for use in future Wasm projects.

We then learned how to publish MQTT messages from inside a Wasm app, by implementing a dashboard able to toggle lights that were represented by an LED.

This book's task was to bring you closer to programming microcontrollers and Wasm and to teach you how to implement small projects with little code and—hopefully—a lot of fun. You have now learned everything you need to go on and realize your own project ideas.

Questions

1. Why is validating credentials inside Wasm code not secure?
2. What are the alternatives to validating credentials inside Wasm code?

Appendix – "Go"ing Ahead

In this appendix section, we are going to have a look at some concepts of the Go programming language that have not been explained in the previous chapters.

The following topics are covered:

- Blocking a goroutine
- Finding heap allocations

Blocking a goroutine

Blocking a goroutine can be important. The easiest example of this is the `main` function. If we have no blocking call or loop inside the `main` function, the program terminates and restarts. In most cases, we do not want a program to terminate, as we might want to wait for a signal on any input that could trigger any further action in the code.

Let's now look at some possibilities for blocking a goroutine. Blocking a goroutine is sometimes necessary in order to gain time to let a scheduler work on other goroutines.

Reading from a channel

A very common way to block a goroutine is to read from a channel. Reading from a channel blocks the goroutine until a value can be read. This is illustrated in the following code example:

```
func main() {
    blocker := make(chan bool, 1)
    <-blocker
    println("this gets never printed")
}
```

A select statement

A `select` statement lets a goroutine wait on multiple operations. The syntax is similar to the syntax of a `switch` statement. The following code example implements a `select` statement that blocks until one of two cases can run:

```go
func main() {
    resultChannel := make(chan bool)
    errChannel := make(chan error)

    select {
    case result := <-resultChannel:
        println(result)
    case err := <-errChannel:
        println(err.Error())
    }
}
```

> **Note**
>
> If both cases happen to be ready at the same time, the `select` statement chooses a random case.

We sometimes have cases where our `main` function should do nothing while other goroutines are waiting for incoming messages to handle them. In such cases, we can make use of an empty `select` statement that blocks indefinitely. An example of this can be seen in the following code snippet:

```go
func main() {
    select {}
    println("this gets never printed")
}
```

Sleeping is a blocking call

In some cases, we only want to gain some time for a scheduler to work on another goroutine. In such cases, we can use `time.Sleep()` in order to sleep for a brief amount of time and then continue to work on our current goroutine. This could look like the following code example:

```go
func main() {
    for {
```

```
        println("do some work")
        time.Sleep(50 * time.Millisecond)
    }
}
```

We have learned some different ways to block a goroutine, so let's now learn a bit about allocations.

Finding heap allocations

The TinyGo compiler toolchain tries to optimize code in such a way that no heap allocations are left in the result, but some allocations cannot be optimized. Is there a way to know which those allocations are? Yes! We can deactivate the **garbage collector** (**GC**) by passing a flag to the `build` and `flash` commands.

When the GC is deactivated, the compilation process is going to fail and throws an error, which points to the line of code that caused a heap allocation. Let's check out the following code example, which causes a heap allocation:

```
package main

var myString *string

func main() {
        value := "my value"
        myString = &value
}
```

When building this program, we will have the GC deactivated with the following command:

```
tinygo build -o Appendix/allocations/hex --gc=none
--target=arduino Appendix/allocations/main.go
```

This is going to throw the following error:

```
/tmp/tinygo589978451/main.o: In function `main.main':
/home/tobias/go/src/github.com/PacktPublishing/Creative-DIY-
Microcontroller-Projects-with-TinyGo-and-WebAssembly/blob/
master/Appendix/allocations/main.go:6: undefined reference to
`runtime.alloc'
collect2: Error: ld returned 1 as End-Status
error: failed to link /tmp/tinygo589978451/main: exit status 1
```

Storing a pointer to a value in a global object results in a heap allocation. How could we improve the program to not allocate heap memory? We could simply omit using a pointer here. Check out the following example:

```
package main

var myString string

func main() {
        value := "my value"
        myString = value
}
```

We can now try to build the program again, using the following command:

```
tinygo build -o Appendix/allocations/hex --gc=none
--target=arduino Appendix/allocations/main.go
```

This command is going to create the output file and will not throw any errors.

If you want to check out which operations cause **heap allocations** and which do not, check out the following link:

```
https://tinygo.org/compiler-internals/heap-allocation/
```

If you want to gain a better understanding of the **heap**, check out the following link:

```
https://medium.com/eureka-engineering/understanding-
allocations-in-go-stack-heap-memory-9a2631b5035d
```

Assessments

Chapter 1

1. The `tinygo info` command.

2. The `tinygo flash` command.

3. The Arduino UNO has a clock speed of 16 MHz. Blinking at 16 MHz is extremely fast and we wouldn't be able to see it. That is why we set the LED to go on and off for a number of milliseconds.

4. You can find the solution in the code repository: `https://github.com/PacktPublishing/Creative-DIY-Microcontroller-Projects-with-TinyGo-and-WebAssembly/blob/master/Chapter01/blink-sos`

Chapter 2

1. We do this in order to prevent the LED from being damaged. Most LEDs in Arduino Starter Kits or similar kits work with voltages below 5V. Driving them with 5V could permanently damage them.

2. Either by using an external pull-up (or pull-down) resistor or by using a built-in resistor.

3. It needs to sleep in order to give the scheduler time to run the goroutine.

4. You can find a solution for this question in the GitHub repository in the `Chapter02` folder under `traffic-lights-blink`.

Chapter 3

1. The key 3 is in row 0 column 2, so the coordinates are 0,2.

2. You can find the solution at the following link: `https://github.com/PacktPublishing/Creative-DIY-Microcontroller-Projects-with-TinyGo-and-WebAssembly/blob/master/Chapter03/safety-lock-keypad-check-key/main.go`

Chapter 4

1. Turning these sensors off saves energy and extends the lifetime of the water level sensors as it slows down corrosion.

2. The circuit is closed when the signal is high in the Signal (In) port.

Chapter 5

1. The 5V pin is deactivated by default. It requires soldering to activate it. Alternatively, the VIN pin could be used when the Arduino is powered through the USB port.

2. `pulseLength` holds the time from sending out the pulse until returning it. So the pulse traveled the distance twice. That is why we have to divide `pulseLength` by 2.

3. Find the solution here: `https://github.com/PacktPublishing/Creative-DIY-Microcontroller-Projects-with-TinyGo-and-WebAssembly/blob/master/Chapter05/touchless-handwash-timer-120seconds/main.go`

Chapter 6

1. An I2C message contains the address of the device the message is dedicated to.

2. The CS pin is being used to signal that a message is dedicated to a specific device as the CS pin is directly connected to the device.

Chapter 7

1. To ensure that a message gets delivered, we need to use QOS level 1 or level 2 as level 0 is a fire-and-forget approach.

2. Yes, none, one, or many clients can subscribe to a topic.

Chapter 8

1. Validating credentials inside the Wasm code is not secure because the Wasm binary is being delivered to the client. The Wasm binary can then be decompiled and the credentials can be extracted.

2. We can always use any kind of authorization service. In general, the credentials should not be validated inside the client logic, but on any other service.

Afterword

Writing this book was great fun. I really enjoyed digging deeper than ever before into TinyGo. I found some problems along the way, created issues, contributed some drivers, and learned a lot. As a consequence of writing this book two new libraries have been created, TinyDom and TinySocket, which can be found on GitHub.

I want to use this opportunity to thank all the people that helped me write this book. First of all there is Alok Dhuri, who found me in the wilds of the internet and gave me the opportunity to write this book. He also provided lots of genius ideas, which have been implemented in some of the projects.

Also, I want to thank Nitee Shetty and Tiksha Abhimanyu Lad for helping me find a good flow in the book by providing tons of feedback on the style, and they also asked tons of good questions in order to get the best out of me.

Also, thanks to everyone else on the Packt team for helping me to get this book done!

And, of course, I want to take the chance to thank Enrico von Otte and Johannes Kolata for instantly accepting my request to be the technical reviewers of the book.

Special thanks also go to Florian Puehl for explaining contract details to me, as well as Martin Nahrgang for helping in terms of PR topics.

And what would an expression of gratitude be without thanking my parents, Carsten and Diana Theel, for raising me?

Other Books You May Enjoy

If you enjoyed this book, you may be interested in these other books by Packt:

Mastering Go – Second Edition

Mihalis Tsoukalos

ISBN: 978-1-83855-933-5

- Clear guidance on using Go for production systems
- Detailed explanations of how Go internals work, the design choices behind the language, and how to optimize your Go code
- A full guide to all Go data types, composite types, and data structures
- Master packages, reflection, and interfaces for effective Go programming
- Build high-performance systems networking code, including server and client-side applications
- Interface with other systems using WebAssembly, JSON, and gRPC
- Write reliable, high-performance concurrent code
- Build machine learning systems in Go, from simple statistical regression to complex neural networks

Hands-On Software Engineering with Golang

Achilleas Anagnostopoulos

ISBN: 978-1-83855-449-1

- Understand different stages of the software development life cycle and the role of a software engineer
- Create APIs using gRPC and leverage the middleware offered by the gRPC ecosystem
- Discover various approaches to managing package dependencies for your projects
- Build an end-to-end project from scratch and explore different strategies for scaling it
- Develop a graph processing system and extend it to run in a distributed manner
- Deploy Go services on Kubernetes and monitor their health using Prometheus

Packt is searching for authors like you

If you're interested in becoming an author for Packt, please visit authors.packtpub.com and apply today. We have worked with thousands of developers and tech professionals, just like you, to help them share their insight with the global tech community. You can make a general application, apply for a specific hot topic that we are recruiting an author for, or submit your own idea.

Leave a review - let other readers know what you think

Please share your thoughts on this book with others by leaving a review on the site that you bought it from. If you purchased the book from Amazon, please leave us an honest review on this book's Amazon page. This is vital so that other potential readers can see and use your unbiased opinion to make purchasing decisions, we can understand what our customers think about our products, and our authors can see your feedback on the title that they have worked with Packt to create. It will only take a few minutes of your time, but is valuable to other potential customers, our authors, and Packt. Thank you!

Index

www.ingramcontent.com/pod-product-compliance
Lightning Source LLC
Chambersburg PA
CBHW062105050326
40690CB00016B/3210